A PHOTOGRAPHIC FIELD GUIDE TO THE
BIRDS
OF
SRI LANKA

First published in the United Kingdom in 2017 by John Beaufoy Publishing Ltd
11 Blenheim Court, 316 Woodstock Road, Oxford OX2 7NS, England
www.johnbeaufoy.com

10 9 8 7 6 5 4 3 2 1

Photo credits
Front cover: Black Bittern © Nirosh Perera; **Spine**: Ceylon Hill-myna © Ajith Ratnayaka; **Back cover**: Plum-headed Parakeet © Nirosh Perera, author photo © Ajith Ratnayaka; **Title page**: Layard's Parakeet © Gehan de Silva Wijeyeratne; **Contents page**: Common Kingfisher © Gehan de Silva Wijeyeratne.
Main descriptions: Photos are denoted by a page number followed by t (top), b (bottom), l (left), r (right), c (centre), i (inset). All photos taken by Gehan de Silva Wijeyeratne except:
Mohammed Abidally 183l; **Duha Alhashimi** 205bl, 248tr, 248br; **Amar-Singh HSS** 101br, 263tl, 263bl, 268t; **Steven An** 109tr, 109br; **Mike Barth** 37tl, 37bl, 58tl, 58bl, 88bl, 106tl, 121tr, 123tr, 125tr, 125br, 136tl, 136bl, 229tl, 229bl; **Peter Beesely** 83tr, 83br, 111br; **Julian Bell** 42tr, 42br; **Biju P.B.** 171r, 247r; **Tamsin Carlisle** 256tl; **Richard Chandler** 65tr, 65br, 109l, 119tr, 125bl/r, 130tr, 134tl, 134bl; **Richard Dawkins** 103br; **Alain de Broyer** 36l; **Atul Dhamankar** 57tl, 67bl; **Wasantha P Dissanayake** 249l; **Keith Dover** 43tl, 43bl, 45tl, 45bl, 48l, 66tr, 149r, 168r, 201tl; **Athula Edirisinghe** 235tr, 258tr; **Graham Ekins** 37tr, 37br, 39tr, 39br, 47tr, 54tl, 56bl, 79bl, 80br/r, 80 r/l, 83tl, 83bl, 121br, 128l, 142tl, 142tr, 153tl, 160r, 173r, 198r, 202r, 206l, 206br, 225tr, 225br, 244r, 260r; **Hanne & Jens Eriksen** 44tl, 44bl, 86tr, 86br (both), 97tl, 125tl, 125bl/l, 137tl, 137bl, 144tl, 144bl, 144tr, 144br, 146br, 244l, 266l, 266br; **Edmund Fellowes** 164tl; **Con Foley** 42l, 52r, 56tl, 59br, 72l, 79tl, 84br, 92l, 98tl, 123tl, 131bl, 150br, 159l, 162r, 163tr/r, 222bl, 224tl, 261i; **Prasad Ganpule** 215tr (both); **Dusty Gedge** 127br; **Clive Griffin** 47br; **Chitral Jayatilake** 148l; **Matt Jones** 38l, 38br, 39tl, 40tr, 40br, 41tl, 41bl, 41r, 44br, 112tl, 120bl, 127tr; **Sunny Joseph** 262tr, 262br; **Lee Tiah Kee** 54br, 57bl, 72r, 78tl, 78bl, 79b far left, 86l, 98bl, 107br, 110tr, 110br, 119tl, 122tr, 131tl, 142br, 163r/l, 180tr, 185bl, 206br, 224bl, 267tl, 267bl; **Rohana Kodikara** 113bl/l; **Pankaj Maheria** 77r, 91r, 256bl; **Manjula Manthur** 66tl, 67tl, 78br, 80r/r, 92tr, 107tr, 122tl, 126tl, 126bl, 126br, 140br, 141tr, 150tr, 184l, 195r, 197tl, 197bl, 201bl, 202l, 205tl, 208bl, 227r, 247tl, 248l, 255r, 256r, 257r; **Ron Marshall** 111tr, 242l; **Chrys Mellor** 35, 36tr, 36br, 75bl, 132l, 132r, 133tl, 133bl, 133cr; **Andrew Moon** 104br, 106bl; **Alan Owyong** 185tl, 222tl; **Mike Parker** 121tr; **Nirosh Perera** 165r/r; 189i, 222tr, 222br, 262l; **Yoav Periman** 81r, 88tl; **Jasbir S Randhawa** 257l; **Ajith Ratnayaka** 7 (2nd from top), 27 (tl), 29 (2nd row from top, right), 63r, 64r, 73r, 74bl, 79r (both), 90r (both), 92 br (both), 93tl, 93bl, 104tl, 117tr, 130l, 137tr, 151tr, 153bl, 155r, 161l, 161r (both), 162l (both), 167bl, 167r, 183r, 189r, 190l, 191tr, 191br, 194l, 203tl, 203bl, 207r (inset), 208tl, 209tl, 211l, 221tl, 228tr, 228br, 230l, 239tl, 243tr, 253l, 254bl, 259tl, 259tr, 259br, 261l, 261r, 263tr, 265tl, 265bl/r, 271tr/r; **John Richardson** 103tr; **Roger Riddington** 123br, 134br, 164bl, 245tr, 245br; **Saurabh Sawant** 80l, 163tl, 163bl; **James Sellen** 87tl, 90tl, 90bl (both), 97bl; **Sumit Sengupta** 66bl, 174r, 247bl, 266tr; **Ramki Sreenivasan** 149tl, 149bl; **Prashanth Srivastava** 82tr, 82br, 87r, 88r, 96tr, 215br, 226l; **Richard Stonier** 38tr, 204tl, 204bl; **Tom Tams** 59tr, 60tr, 60br, 67br, 76l, 77tl, 77bl, 84l, 98tl, 98br, 108r, 112bl, 114tr, 120tl, 124tl, 124bl, 130br, 139tl, 139bl, 164r, 166l, 168l, 170l, 172l, 172r, 175l, 189l, 196r, 201br, 218l, 219l, 228tl, 228bl, 231l, 231r, 233tl, 233bl, 233r, 243br, 270r; **Tom Tarrant** 106tr, 106br; **Jyotendra Thakuri** 94r; **Thet Zaw Naing** 89l; **Fran Trabalon** 174l, 175r, 268b; **Veer Vaibhav Mishra** 173t; **Rajiv Welikala** 28 (5th row from top), 28 (6th row from top, left), 28 (8th row from top), 29 (2nd row from top, left), 34r, 50r, 55bl, 73tl, 103cl, 140tl, 179bl, 221tr, 225l, 234l, 236l, 241tl, 251b, 262b; **David Williams** 200l, 232tr, 232br; **Cherry Wong** 44 tr, 48tr, 48br, 49 (all), 133br; **Michelle & Peter Wong** 63bl, 66br, 96br, 99bl, 119bl, 122bl, 122br, 123bl, 142bl, 154r, 176l, 180br, 241tr, 241bl, 242tr, 242br, 245l, 246br.

Great care has been taken to maintain the accuracy of the information contained in this work. However, neither the publishers nor the author can be held responsible for any consequences arising from the use of the information contained therein.

ISBN 978-1-909612-83-9

Edited by Krystyna Mayer
Designed by Gulmohur
Cartography by Gulmohur
Project management by Rosemary Wilkinson

Printed and bound in Malaysia by Times Offset (M) Sdn. Bhd.

A PHOTOGRAPHIC FIELD GUIDE TO THE
BIRDS
OF
SRI LANKA

Gehan de Silva Wijeyeratne

JOHN BEAUFOY PUBLISHING

Contents

Introduction

This book covers all bird species recorded in Sri Lanka. It builds on the author's earlier title published as *A Naturalist's Guide to the Birds of Sri Lanka*. Both titles are intended to be portable books for field use. As many of the species are found in India as well as Sri Lanka, this book has a wider application beyond the island.

The text on the birds is aimed primarily at eliminating some of the confusion that may arise in identification in the field because of often varying plumages in the same species, or some species looking similar to each other. Readers are encouraged to consult other books to develop their field craft, as each book has different nuances. Other useful books include *A Guide to the Birds of Sri Lanka* by G. M. Henry, which is a very useful reference volume that was once the standard guide to Sri Lanka's birds, and *A Field Guide to the Birds of Sri Lanka* by John Harrison and Tim Worfolk. A more compact field guide in the Helm Field Guide series is *Birds of Sri Lanka* by Deepal Warakagoda, Carol Inskipp, Tim Inskipp and Richard Grimmett, adapted from the landmark *Birds of the Indian Subcontinent*.

Bird Voices

Calls and songs are very important in distinguishing species, and brief descriptions of them are given for the birds that are vocal. A bird can have a range of vocalizations, depending on its age and sex, and any inter-specific and intra-specific interaction. I have tried to describe representative calls – though it is always a struggle to transcribe vocalizations and everyone has their own method of doing so. Voices are only described for birds that are generally vocal and where the vocalizations are useful in field identification. In some species the vocalizations are in fact the crucial clue to recording a bird's presence or absence at a site. Some birds tend to be generally quiet (for example various waterbirds and seabirds), so I have not described their voices. Many of these birds do utter grunts and squawks occasionally, for example at nesting colonies, but these have virtually no value in field identification.

CDs to the songs and calls of Sri Lankan birds have been produced by Deepal Warakagoda and a few other sound recordists. These are useful for birdwatchers who wish to develop their field craft further. There are also many resources on the internet for learning bird calls. One of the best is www.xeno-canto. org, which was accessed to complement my own sound recordings when writing this book. Others include http://avocet.zoology.msu.edu. The best way to learn birds' vocalizations is to listen carefully in the field and refresh your memory by listening to recordings.

The Photographs

Many of the featured breeding birds were photographed in Sri Lanka by the author and Sri Lankan photographers. Almost all the Highly Scarce Migrants and Vagrants were sourced from photographers who photographed them in areas where they are regular visitors or summer breeders. These species have no differences diagnosable in the field with the species or subspecies seen in Sri Lanka.

Pictures contributed by other photographers are individually credited, and I am very grateful to the support given by many photographers from all over the world. The sourcing of images was made easier thanks to online image databases, especially the sites maintained by Bird Guides (www.birdguides.com) and the Oriental Bird Club (orientalbirdimages.org). With regard to the latter, I must especially mention Krys Kazmierczak, who has devoted thousands of hours to making the site a success.

Plumage

Where relevant the images of birds are captioned for plumage. Birds can differ in the plumage for four main reasons: 1 Age; 2 Sex; 3 Breeding or Non-breeding; 4 Summer or Winter (related to the Breeding or Non-breeding category for migrants).

All birds will have a juvenile plumage after fledging. In some birds this may only last a few weeks or months. Juveniles often show pale edges in their wing feathers. They may also show other features, such as spotting, streaking and pale gape lines. They are usually duller than adults and with some birds (e.g. cormorants) the juvenile may have pale underparts. With some groups of birds such as gulls and raptors young birds may take two to four years to reach adult plumage. Such birds are often referred to as immatures. In advanced family monographs or technical papers, they may be referred to by the calendar year. A bird is deemed to have its birthday on 1 January. Thus a gull that hatched in summer and is in post juvenile moult in December will be classified as a 1cy bird for the first calendar year; 2cy and 3cy are second and third calendar year birds respectively. As this field guide is intended for beginners as well, I have not used this notation. 'Immature' will indicate a bird that has a complex moult spanning a year or more to attain adult plumage.

Some species have distinct adult plumages or sizes for the adult male and female. This is known as sexual dimorphism. Sunbirds are a good example with the male usually being brighter and more showy. In the case of raptors, the female may also be bigger. This helps the two sexes to avoid competition for prey in their territory.

Many birds also undergo a short-lived change in plumage during the breeding season. With resident birds such as egret and herons, bare parts on the face and the legs can assume bright colours. Egrets also develop plumes on back of the head. In passerines the colours of the male may be brighter in breeding plumage. In ducks, males are brightly coloured when they are ready to breed and they undergo a post breeding moult where they shed the colours and look like the drab female.

Wading birds show a pronounced change between their breeding plumage which can have hues of chestnut to the greyer, duller colours of non-breeding. As this coincides with summer and winter, the latter is used to help with the additional distinction of which birds are migratory and are winter visitors to Sri Lanka.

If no plumage type is indicated it would mean the adult bird has no pronounced sexual dimorphism or breeding season changes.

Male Loten's Sunbird

Female Loten's Sunbird

Immature Brown-headed Gull (first winter)

Adult Brown-headed Gull in winter

Climatic Zones & Monsoons

The topography of Sri Lanka comprises lowlands along the perimeter, which in the southern half give rise within a short distance to the central hills, rising to above 2,400m in altitude. A careful examination of the topography reveals that the island can be divided into three peneplains, or steps, first described by the Canadian geologist Frank Dawson Adams in 1929. The lowest peneplain is from 0 to 30m, the second rises to 480m and the third rises to 1,800m.

Sri Lanka can be broadly divided into four regions (Low Country Wet Zone, Hill Zone, Low Country Dry Zone and Intermediate Zone) resulting from the interactions of rainfall and topography. Rainfall is affected by monsoonal changes that bring rain during two monsoons; the south-west monsoon (May–August) and the north-east monsoon (October–January). Their precipitation is heavily influenced by the central hills. These monsoons deposit rain across Sri Lanka and contribute to the demarcation of climatic regimes.

LOW COUNTRY WET ZONE

The humid, lowland wet zone in the South-west of Sri Lanka does not have marked seasons, being fed by both the south-west and north-east monsoons. The Low Country Wet Zone receives 200–500cm of rain from the south-west monsoon, and afternoon showers from the north-east monsoon. Humidity is high, rarely dropping below 97 per cent, while temperatures range from 27 to 31° C over the year.

The Low Country Wet Zone is the most densely populated area in Sri Lanka. The coast is well settled, while the interior has Coconut *Cocos nucifera* and rubber *Hevea* sp. plantations, some rice (paddy) cultivation and small industries. Remnants of rainforests and tropical moist forests exist precariously in some parts of the interior, under pressure from an expanding population. It is in these forests that most of the endemics that are a draw to ecotourists can be found.

Annaiwilundawa, one of many man-made lakes in the dry zone

Tea estates adjoin the cloud forest of Horton Plains National Park

HILL ZONE

The mountainous interior lies within the wet zone and rises to more than 2,400m. Rainfall is generally well distributed, except in the Uva Province, which gets very little rainfall in June–September.

Temperatures are cooler than in the lowlands and can vary from chilly in the mornings to warm by noon. In the mid-elevations such as the area around Kandy, the temperature varies between 17 and 31° C during the year. Temperature variations during a 24-hour cycle are, however, far less varied. The mountains are cooler, within a band of 14–32° C during the year. There may be frost in the higher hills in December and January, when night-time temperatures fall below zero.

The central Hill Zone is intensely planted with tea, but has small areas of remnant forest and open grassland.

LOW COUNTRY DRY ZONE

The rest of the country, three-quarters of Sri Lanka's land area, consists of the dry zone of the northern, southern and eastern plains. These regions receive 60–190cm of rain each year, mainly from the north-east monsoon. The dry zone is further divided into the arid zones of the North-west and South-east, which receive less than 60cm of rain as these areas are not in the direct path of the monsoonal rains.

The coastal plains in the Southern Province, where Yala and Bundala National Parks are located, and the North Central Province where the cultural sites are located, are dry and hot. Much of the dry zone is under rice and other field crops, irrigated by vast man-made lakes (the tanks, or *wewas*), many of which are centuries old, and were built by royal decree to capture the scarce rainfall in these areas. Once the 'Granary of the East', exporting rice as far as China and Burma, wars and invasions, malaria and other diseases laid

Yala has plains interspersed with scrub and monsoon forest

waste to vast areas of the Low Country Dry Zone. The once-bountiful rice plains were reclaimed by scrub jungle, the haunt of elephants, bears and leopards. Since independence in 1948, successive governments have vigorously promoted colonization and resettlement of these areas. Sandy beaches fringe the coastline, and it is always possible to find a beach that is away from the path of the prevalent monsoon.

INTERMEDIATE ZONE

The Intermediate Zone is a transition zone between the dry and wet zones. Recent rainfall data shows that the wet zone with the highest precipitation is smaller than shown in maps of a few decades ago.

Birdwatching in Sri Lanka

This section provides a brief overview of birdwatching in Sri Lanka. For brevity, these are some of the highlights of birdwatching in Sri Lanka.

- There are 34 endemic birds in Sri Lanka – a high density of endemics per unit area.
- Birds are surprisingly tame whether in a national park or in an urban setting.
- Even birds familiar worldwide, such as waders, present better photographic opportunities in Sri Lanka than elsewhere, as in many sites the birds can be approached very closely.
- On this moderately sized island, the birding sites are highly varied, including lush lowland rainforests, cloud forests, elephant and leopard-rich dry lowlands, wetlands and coastlines.
- There is a good road and accommodation infrastructure, and a mature tourism industry. With English widely spoken in cities and even in remote areas, you are likely to always find someone who can understand a bit of English.
- A number of specialized wildlife and birding tour companies cater to both local and foreign birders.
- There is a good suite of natural history publications relevant to Sri Lankan birds.
- A pdf of the article 'Why Sri Lanka is Super-rich for Wildlife' is available on the Internet. The article uses a number of graphics to explain its topic

Habitats & Top Sites

For a moderately sized island, Sri Lanka offers a variety of habitats. It is an oceanic island with a continental shelf coming close to it in the South at Dondra and also near the Kalpitiya Peninsula. Submarine canyons cut into the Trincomalee harbour in the North-east. A chain of partially submerged islands (Adam's Bridge) connect it to the Indian mainland. The entire coastline offers suitable habitat for seabirds, with the western coast especially off Kalpitiya being the richest. Estuaries, mudflats, mangroves and man-made salt pans off the North-west and South-east are especially rich habitats for waders. Mannar Island and the neighbouring mainland have internationally significant numbers of migrant waders and ducks. The island is dotted with more than 2,000 man-made freshwater lakes in the dry zone, which are rich in waterbirds. As many as 103 river systems create a rich network of aquatic habitats that are further enriched by paddy fields, which form an important artificial system of wetlands. The dry zone is characterized by grassland, thorn scrub and wooded sections where the soil and rainfall support them, or where 'gallery forests' remain along watercourses. The lowlands in the South-west originally held rainforests, though much of these have been lost to cultivation. At higher elevations, the highlands in the wet zone hold cloud forest, now a tiny remnant of what once existed. The highlands are interspersed with patana grassland.

In the context of visiting foreign birders, the key targets would be the endemics that are found in the wet zone. For this, a good lowland rainforest site such as Sinharaja and a montane site such as Horton Plains National Park are essential places to visit. Birders in search of the endemics should also fit in Kithulgala Rainforest to maximize their chances of seeing them. Given that Sri Lanka is the best for big game safaris outside Africa, a visit to a national park such as Yala is also recommended. The following are the main sites visited by birders on a two-week tour to Sri Lanka.

LOW COUNTRY WET ZONE	
Talangama Wetland	In Colombo's suburbs, visited on arrival or before departure. Good range of wetland birds, often at close quarters.
Bodhinagala	Optional: a small but rich patch of rainforest about an hour and a half from Colombo.
Sinharaja	The most important site. All but a handful of endemic birds can be seen here.
Morapitiya	Optional: Morapitiya on the way to Sinharaja can at times be rewarding. Check the state of the access road, as this is highly variable.
Kithulgala	A rainforest in the mid-hills that has the mix of bird species found in Sinharaja. An essential second stop for those wanting to see all the endemics.
HILL ZONE	
Horton Plains National Park	This, together with sites such as Hakgala Botanical Gardens, is essential for some of the montane endemics such as the Ceylon Whistling-thrush. Birders also visit Victoria Park in Nuwara Eliya town.
LOW COUNTRY DRY ZONE (SOUTH)	
Yala, Bundala, Palatupana	Yala for leopards, elephants and Sloth Bears. Uda Walawe for elephants. Both parks are good for a variety of dry zone birds. Palatupana Salt Pans and Bundala National Park offer good views of migrant waders and other birds.
LOW COUNTRY DRY ZONE (NORTH-CENTRAL)	
Mannar Island	For migrant waders, waterfowl, gulls and Deccan Plateau residents that are rare or absent further south. Generally of more interest to resident birders, unless a visitor has a special interest in waders or gulls.
Minneriya and Kaudulla National Parks	For the Elephant Gathering, which peaks in August and September.

Talangama Wetland

This wetland on the outskirts of Colombo is bordered by motorable roads, which make access easy for birdwatchers. The complex of ponds, canals and paddy fields makes it a rich and varied wetland site. Colombo as a capital is blessed to have such a suburban jewel.

Birdwatching Highlights
More than a hundred bird species have been recorded here. Key species are the migrant Black and Yellow Bitterns, which augment the local population, and the Watercock. Talangama is also good for the most common butterflies and dragonflies. More than 30 dragonfly species are found here.

Getting There
From Colombo get to Wewa Para (Lake Road) via Akuregoda Road or Wickramasinghapura

Talangama Wetland

Road, both of which are off the Pannipitiya Road, a few kilometres from the Parliament. Free access on public roads.

Accommodation
Villa Talangama overlooks one of the best stretches of wetland. City hotels in Colombo are only 30–45 minutes away.

BODHINAGALA

Bodhinagala is a relatively small tract of secondary lowland rainforest, with a Buddhist hermitage located centrally reached through a forest path. It is surprisingly rich floristically and holds a number of endemic fauna within relatively easy reach of the commercial capital of Colombo.

Bodhinagala

Birdwatching Highlights
Bodhinagala's claim to fame with birders is as a site for the endemic Green-billed Coucal. It also contains a number of other endemics such as the Ceylon Spurfowl, Yellow-fronted Barbet, Ceylon Small Barbet, Black-capped Bulbul and Spot-winged Thrush, and subcontinental endemics such as the Ceylon Frogmouth and Malabar Trogon. Butterflies include the Tawny Rajah.

Getting there
The turn-off to Bodhinagala is just before the 29 km post on the A8 (Ratnapura Road).

Accommodation
Using private transport, the site can be visited as a half-day trip from Colombo, which has a wide choice of accommodation.

SINHARAJA RAINFOREST

The Sinharaja Man and Biosphere Reserve was declared a UNESCO World Heritage Site in 1988. It is arguably the most important biodiversity site in Sri Lanka and is also internationally important for tropical biodiversity. It is famous for its bird waves (see pages 21 and 230). Sinharaja comprises lowland and submontane wet evergreen forests with submontane Patana grasslands in the east. A staggering 64 per cent of the tree species are endemic to Sri Lanka. The lower slopes and valleys have remnant dipterocarpus forest, with the middle and higher slopes characterized by trees of the genus Mesua. Orchids (Orchidaceae spp.) and pitcher plants (Nepenthaceae spp.) are common in nutrient-poor soils.

Sinharaja Rainforest

Birdwatching Highlights
Endemic birds include the Ceylon Spurfowl, Ceylon Junglefowl, Ceylon Wood Pigeon,

Ceylon Hanging-parrot, Layard's Parakeet, Red-faced Malkoha, Green-billed Coucal, Serendib Scops-owl, Chestnut-backed Owlet, Ceylon Grey Hornbill, Yellow-fronted Barbet, Ceylon Small Barbet, Greater Flameback, Black-capped Bulbul, Spot-winged Thrush, Ceylon Rufous and Brown-capped Babblers, Ashy-headed Laughingthrush, Ceylon Blue Magpie, White-faced Starling, Ceylon Hill Myna, Ceylon Scaly Thrush, Ceylon Scimitar Babbler and Ceylon Crested Drongo. Indian subcontinental endemics include the Malabar Trogon and Ceylon Frogmouth.

Getting There
Access is possible from Pitadeniya, but not practical for most visitors. Motorable access is to Kudawa via Ratnapura or via Buluthota Pass from Yala. From the coast via Katukurunda Junction, Agalawatta and Kalawana. If using the southern expressway take the Dondangoda exit which is near Mathugama.

Accommodation
Boulder Garden at Kalawana and Rainforest Edge at Veddagala provide the nearest star-quality accommodation. Serious birders can look at Martin's and Blue Magpie Lodge, near the reserve.

Morapitiya Rainforest
A lowland wet zone forest which is floristically and faunistically similar to Sinharaja. It would have been contiguous with Sinharaja perhaps into the 19th century after which the island experienced a population explosion. Like Sinharaja, Morapitiya is also under the jurisdiction of the Forest Department. It is a little unusual in that there is a public road that runs into a village in its interior. Because of the regular passage of people, the bird waves have become habituated to people. One drawback is the absence of any facilities for visitors and the condition of the access road is variable.

Birdwatching Highlights
The secondary forest edges have a high density of Serendib Scops-owls and on some evening visits I have heard several territory-holding birds. It is a good site for the Green-billed Coucal, Red-faced Malkoha and Ceylon Crested Drongo.

Morapitiya Rainforest

Getting there

The access road is a turn-off at Athwelthota on the way to the more popular Sinharaja. The first few kilometres of the road convert from village gardens to a mix of village gardens and rainforest.

KITHULGALA (KELANI VALLEY) RAINFOREST

Kithulgala (Kelani Valley Forest Reserve) was established to protect the watershed of the Kelani River. It is home to much of Sri Lanka's endemic fauna and flora.

Birdwatching Highlights

A good number of endemic birds, including the Spot-winged Thrush, Green-billed Coucal, Red-faced Malkoha, Ceylon Grey Hornbill, Yellow-fronted Barbet, Ceylon Spurfowl, Ceylon Rufous Babbler, Ceylon Scimitar Babbler and Ceylon Frogmouth, are found here.

Kithulgala (Kelani Valley) Rainforest

Getting There

The Kithulgala Rest House is just after the 37 km post on the A7. Take the ferry across the river and access the forest using the village trails.

HORTON PLAINS NATIONAL PARK

Sri Lanka's second and third highest peaks, Kirigalpotta (2,395 m/7,860 ft) and Thotupola Kanda (2,357 m/7,733 ft), are found here. Three important rivers, the Mahaveli, Kelani and Walawe, originate from Horton Plains.

Horton Plains National Park

Birdwatching Highlights
Endemic birds include the Ceylon Whistling Thrush, Ceylon Hill White-eye, Ceylon Wood Pigeon and
Dusky-blue Flycatcher.

Getting There
Starting from Nuwara Eliya, about 6 km (3¾ miles) from town on the A7, is a left turn towards Ambewala
and Pattipola. This continues to the park. From Haputale, take the road via Ohiya.

Accommodation
Nuwara Eliya has a wide choice of accommodation.

YALA (RUHUNA) NATIONAL PARK

Yala is undoubtedly Sri Lanka's most visited national park, and the best in Sri Lanka for viewing a wide
diversity of animals. It is a wonderful place, with a spectrum of habitats from scrub jungle, lakes and
brackish lagoons to riverine areas. The park is divided into five blocks, of which Block 1 (Yala West) is
open to the public. Yala may be closed between 1 September and 15 October. The flora is typical of dry
monsoon forest vegetation in the southern belt.

Birdwatching Highlights
Endemic birds include the Ceylon Junglefowl, Brown-capped Babbler, Ceylon Wood-Shrike and Ceylon
Swallow. The park is also a good place to see dry-zone specialities like Eurasian and Great Thick-knees,
Sirkeer, Blue-faced Malkohas and Malabar Pied Hornbills. It is probably the best place to see the rare Black-
necked Stork. A day's birding in the park during the northern winter can yield a hundred species.

Getting There
About 40 km (25 miles) beyond Hambantota on the A2.

Yala (Ruhuna) National Park

Bundala National Park

Accommodation
Tissamaharama has a broad range of accommodation. Near the turn-off to the park at Kirinda is the Elephant Reach and a few smaller properties. Close to the park, are two top-end safari lodges, the Jetwing Yala and the Chaaya Wild.

Bundala National Park
Bundala National Park is a mix of scrub jungle and sand dunes bordering the sea. Its beaches are important nesting sites for turtles. The lagoons hold good numbers of birds and crocodiles.

Birdwatching Highlights
Endemic birds include the Brown-capped Babbler, Ceylon Wood-Shrike and Ceylon Junglefowl. During the northern winter large numbers of migrants arrive, such as Golden and Kentish Plovers, Large and Lesser Sand Plovers, Marsh Sandpipers and Curlew Sandpipers, Curlew and Common Greenshank. Rarities include the Broad-billed Sandpiper and Red-necked Phalarope.

Getting There
From the A2, at the Weligatta Junction near the 251 km post, take the turn to Bundala Village. The park office and entrance are on this road.

Accommodation
Tissamaharama has a range of places to stay. Alternatively, there is accommodation close to Yala National Park.

Palatupana Salt Pans

PALATUPANA SALT PANS

This is a salt pan administered by the Salt Corporation. For close views of waders, it is probably one of the best sites in the world. Some of the adjoining areas can be seen on the Kirinda- Yala Road. These areas are also good for waders and terns. However, the best views of waders are when you drive in through a gate manned by a security guard into the working salt pans.

Birdwatching Highlights

The salt pans are excellent for waders such as Stints, Curlew and Common Sandpipers, Lesser Sand Plover, Grey Plover, Terek Sandpiper, Black-winged Stilt and star attractions like Broad-billed Sandpiper and Red-necked Phalarope. Between five to ten thousand migrant White-winged Black Terns have been seen to roost here together with small numbers of other migrant terns including Common and Gull-billed Terns. Other water birds include the internationally threatened Spot-billed Pelican and migrant waterfowl like Pintail, Garganey and Shoveler.

Accommodation

As for Yala National Park.

Getting There

On the Kirinda-Yala Road (B499).

MANNAR ISLAND

Mannar Island and the strip on the mainland from around Giant's Tank have become magnets for birders in search of species that are not found regularly in the southern half of the island. These include Deccan avifaunal species such as the Long-tailed or Rufous-rumped Shrike, Black Drongo, Crab Plover and Indian Courser. A few key sites in this area are described below.

Mannar Island

Periyar Kalapuwa (lagoon): A finger of this lagoon crosses the A14, about 4 km (2½ miles) before the Mannar Causeway, near the 78 km post. Look for Garganey, Common Teal and Common Ringed Plover. The seasonal wetland holds thousands of Wigeon and a few hundred Shovellers. The plains are also good for harriers (Circinae spp.).

Mannar Causeway: The star birds here are the Oystercatcher, and Pallas's and Heuglin's Gulls. All three species are rare in the South. The causeway also allows close views of the Whimbrel and Curlew, and at times the Avocet and Crab Plover.

Talaimannar: About a kilometre from the now-defunct Talaimannar Customs post there is a 'fishing port'. Large flocks of gulls gather here, including Heuglin's, Brown-headed and Pallas's Gulls.

Sand Banks (Adam's Bridge): A series of islands form what is known as Adam's Bridge, connecting Talaimannar to Rameswaran in the south-west of India. During the breeding season take care not to disturb the hundreds of nesting terns. Access requires permission.

Accommodation
Mannar has a few small guest houses and the Palmyrah House – a luxury boutique hotel.

MINNERIYA NATIONAL PARK

The Minneriya and Kaudulla National Parks are centred around two large reservoirs (known locally as 'tanks') constructed by King Mahasen in the 3rd century AD. Many centuries ago, these lowlands were farmed for agriculture by an ancient civilization whose mastery of hydraulics was remarkably sophisticated. Today, the ancient reservoirs fill during the north-east monsoon and gradually shrink as the dry season fastens the lowlands in a torpid grip. As the waters recede, lush grassland sprouts attracting elephants in search of food from as far away as the jungles of Wasgomuwa and Trincomalee. The grassland is bordered by monsoon forests, which have some fine stands of old trees. The tall forest is interspersed with a few patches of thorn scrub contributing to a matrix of habitats. During the rainy season the lake becomes inaccessible due to mud and water. People in search of elephants visit Hurulu Reserve, a Forest Department reserve off the A6, near the Habarana Junction.

Minneriya National Park

Birdwatching Highlights
In the forest around the lake, endemic birds that may be seen include the Ceylon Junglefowl, Brown-capped Babbler, Ceylon Grey Hornbill and Black-capped Bulbul. The open areas around the lake are good for raptors including the Brahminy Kite, Grey-headed Fish-eagle and White-bellied Sea Eagle. At the water's edge, Little, Intermediate and Great Egrets can be seen. Large numbers of migrant Whiskered Terns hunt over the water.

Getting There
The Minneriya park entrance is near the 35 km post on the A11 running between Habarana and Polonnaruwa.

Accommodation
There are good hotels at Habarana (the nearest), Giritale, Polonnaruwa, Sigiriya and Kandalama.

Kaudulla National Park
This national park, as the crow flies, is a few kilometres away from Minneriya National Park. It, too, is centred around an ancient reservoir built by King Mahasen and is very similar to Minneriya in terms of its physical structure and the composition of its plants and animals.

Getting There
The turn-off to the park entrance is off the A6 (Habarana-Trincomalee Road) about 15 km (9 miles) from Habarana Junction. The roads to Minneriya and Kaudulla lead off to the east and north respectively from Habarana junction and their respective entrance gates are around 25 km (15 miles) apart, although the two parks have boundaries which are only a few kilometres apart. On a game drive, visitors tend to choose one or the other, depending on conditions for access and where the elephants are gathered.

Birdwatching Highlights and Accommodation
Similar to Minneriya National Park.

Kaudulla National Park

Additional Birdwatching Sites

Uda Walawe National Park

Uda Walawe is a popular national park because of its elephants. The park is a mixture of abandoned Teak plantations, grassland, scrub jungle and riverine 'gallery forest' along the Walawe Ganga and Mau Ara. Uda Walawe is probably the best place to see wild herds of elephants, consisting of tightly-knit family groups of up to four generations of adult and sub-adult females and young. Satinwood, Ebony and Trincomalee Wood trees are present and the river margins are characterised by water-loving Kumbuk trees.

Uda Walawe Reservoir

Birdwatching Highlights

Endemic birds include the Ceylon Junglefowl, Ceylon Spurfowl, Ceylon Grey Hornbill, Ceylon Woodshrike and Ceylon Swallow. In forested areas, Sirkeer and Blue-faced Malkohas are found.

Getting there

The park entrance is on the B427 between Timbolketiya and Tanamanwila, near the 11 km post. From Colombo take the A8 to Ratnapura, A4 to Pelmadulla and A18 to Timbolketiya. It takes around three and a half hours to drive the 180 km (112 miles).

Accommodation

A range of accommodation is available including the Safari Village at Timbolketiya and Centauria Tourist Hotel at Embilipitiya.

Kalametiya Sanctuary

Kalametiya is an extensive area of wetland with brackish lagoons, mangrove swamps, open grassy areas and pockets of scrub jungle. It is a significant site for migrant waders and provides an important refuge (one of the few remaining on the southern coastal strip) for the smaller mammals of Sri Lanka.

Birdwatching Highlights

Almost all of the common wetland birds can be seen here. Sought-after species include Slaty-breasted Crake, Watercock and Black Bittern. During the northern winter, Glossy Ibis may be present with thousands of waders. One of the most reliable sites for Indian Reed-warbler.

Kalametiya Sanctuary

Getting there

There are turn-offs to the sanctuary near the 214 and 218 km posts on the A2 near Hungama.

Accommodation

In Tangalle there is a mix of hotels from the large Tangalle Bay Hotel to small guest houses, which are primarily catering to beach tourists. Most birders stop by on their way to Yala.

ANNAIWILUNDAWA WETLAND

Annaiwilundawa refers to a cluster of freshwater tanks (including the Annaiwilundawa Wewa) that was declared a sanctuary in 1997. The second Ramsar site in Sri Lanka, it is one of the finest wetlands in the island for water birds. Annaiwilundawa is not on the birding tour itineraries for visiting birders but is a worthwhile stop if you are on your way to Mannar or you are staying in Kalpitiya for marine mammals.

Annaiwilundawa

Birdwatching Highlights

Waterfowl include Little Grebe, Lesser Whistling-duck and Cotton Teal. Migrant birds include Pintail, Garganey Common and Pintail Snipe. Large numbers of Asian Openbill and Little Cormorants nest here. Endemics include Ceylon Woodshrike and Ceylon Swallow.

Getting there

At the 91 km post on the A3, 5 km (3 miles) past Arachchikattuwa town, is a turn-off to the left. Approximately 1.2 km (3/4 mile) down this road is Suruwila tank on your left and to your right is the main Annaiwilundawa tank.

Accommodation

Negombo near the 31 km post on the A3 has a wide choice of accommodation. Kalpitiya has accommodation but mainly at the top end.

WILPATTU NATIONAL PARK

Wilpattu National Park comprises a complex of lakes called villus surrounded by grassy plains, set within scrub jungle. The biggest draw here is the leopard. It is also the best place in the country to see Barking Deer also known as Muntjac. The park is just over 1,300 km2 (500 sq miles) and is the largest national park in the country. For comparison, the Yala Protected Area Complex, which comprises a complex of national parks and sanctuaries, is just over 1,500 km2 (580 sq miles). On the east the park is bordered by the sea with the Dutch and Portugal Bays. These bays with their seagrass beds are important for the rarely seen Dugong. Wilpattu is not on the standard itineraries for bird tours as the birds here are covered in other sites such as Yala. However, for those on a mammal tour, especially hoping for Leopard and Sloth Bear, Wilapattu is important.

Birdwatching Highlights

Endemic birds include the Ceylon Junglefowl, Brown-capped Babbler, Ceylon Woodshrike and Black-capped Bulbul in riverine habitats. Good for waders during migrant season with close approaches possible to Pintail Snipe and sandpipers.

Getting there

The turn-off to the Wilpattu National Park is near the 45 km post of the A12. From here, follow the B028, for about 8 km (5 miles).

Wilpattu National Park

Accommodation

On the turn-off to Wilpattu off the A12 (Puttalam to Anuradhapura road) are a few hotels ranging from simple hotel-type accommodation to tented safari lodges off the beaten track. The nearest, for a broad range of accommodation is Anuradhapura, which includes the comfortable Palm Garden Village. With improved roads, Wilpattu is also easily accessible from hotels on the Kalpitiya Peninsula and further afield from Negombo.

Rainforests of Galle

These rainforests are not on the standard birding itineraries because the wet lowland endemics are covered by Sinharaja and Kithulgala. However, for those staying in Galle on holiday, there are three sites which are worth visiting.

KANNELIYA RAINFOREST

Kanneliya is one of the last remaining large tracts of lowland rainforest in Sri Lanka. Its importance is on a par with Sinharaja, with similar fauna and flora. It is a mix of logged secondary and virgin forest.

Kanneliya

Birdwatching Highlights

Oddly lowland endemic birds such as Ceylon Rufous Babbler are missing, although other scarce endemic species such as the Serendib Scops Owl are present. Most of the species described in Sinharaja can be seen here but the birds will not be as habituated.

Getting there

From Galle, take the B129 (Udugama Road) to Udugama. Continue towards Hiniduma and the turn-off to Kanneliya is after the 3 km post on the B429.

Accommodation

Galle between one and a half to two hours drive, has a wide choice of accommodation.

KOTTAWA RAINFOREST AND ARBORETUM

The Kottawa Arboretum is a small tract of lowland secondary rainforest that holds a surprising number of endemic species. It is a part of the Kottawa-Kombala Conservation Forest of 1,800 hectares (4,450 acres).

Kottawa

Birdwatching Highlights

For a small place, this is very good for endemics, which include Spot-winged Thrush, Ceylon Spurfowl, Ceylon Hanging Parrot, Layard's Parakeet, Ceylon Small Barbet, Legge's Flowerpecker, Black-capped Bulbul, Ceylon Grey Hornbill, Chestnut-backed Owlet, Yellow-fronted Barbet and even Ceylon Hill Munia. Hard to get, but found here is the subcontinental endemic, the Sri Lanka Frogmouth.

Getting There
From Galle take the Udugama Road, the B129. This is opposite the turn-off to the Closenberg Hotel and Galle Port (which you pass on the right if you are heading to Matara). Just past the 13 km post on the B129, on the right, is the Kottawa Information Center. Buy your entrance tickets here. Further along the road before the 14 km post are gates to the left and a large yellow signboard "Kottawa Arboretum Wet Evergreen Forest Kottawa Kombala". Enter the forest from here. Follow the wide trail that runs parallel to the road until it rejoins it about 1 or 2 km (about 1 mile) further down.

Accommodation
A wide range of accommodation is available in Galle and elsewhere in the coastal strip.

HIYARE RAINFOREST PARK
(Environment & Biodiversity Study Center & Botanical Garden)

Hiyare is a reservoir bordered by 245 hectares (600 acres) of secondary lowland rainforest. The reservoir was established in 1911 and encompasses 22 hectares (55 acres). It is managed by staff of the Galle Municipality. The Forest Department also has jurisdiction as the reservoir adjoins the Kottawa-Kombala Forest Reserve. This should not be confused with the Kottawa Rainforest & Arboretum, which is a few kilometres further down along the Udugama Road (B 129).

Birdwatching Highlights
Endemics include Ceylon Spurfowl, Brown-capped Babbler, Spot-winged Thrush, Black-capped Bulbul, Ceylon Grey Hornbill and Ceylon Hanging Parrot. Easier to see birds here than in the Kottawa Rainforest.

Accommodation
A wide range of accommodation is available in Galle and elsewhere in the coastal strip.

Hiyare

Itinerary

The itinerary below is typical of that taken by many visiting birders who are trying to maximize the number of species seen while also taking into account the endemics. Remember that the migrants will only be encountered during the migrant season from October to March.

Day 1 Arrive in Sri Lanka and transfer to Colombo for one night. Afternoon, walk in Talangama Wetland for Black and Yellow Bitterns, Watercock and other common sub-continental species.

Day 2 Morning, drive to Bodhinagala Forest Reserve, which is a lowland rainforest closer to Colombo. Look for Ceylon Grey Hornbill, Yellow-fronted Barbet, Ceylon Hanging-parrot, and if lucky the scarce endemic Green-billed Coucal. Thereafter proceed to Sinharaja for three nights.

Days 3–4 Early-morning and late-afternoon walks to Sinharaja Rainforest (a UNESCO World Heritage Site) for lowland endemics and mixed species bird flocks. According to a study of the mixed species bird flocks, on average 42 individual birds occur in a flock, which makes this the world's largest bird wave. The study of the Sinharaja Bird Wave has continued since 1981, and is considered to be the world's longest bird-flock study. Birding highlights include Red-faced Malkoha, Ceylon Blue Magpie, White-faced Starling and Ceylon Scaly Thrush.

Day 5 After breakfast leave for Uda Walawe for one night. Afternoon, game drive at Uda Walawe National Park for dry zone birding and Asian Elephant. Uda Walawe is the only park in the world where a wild elephant sighting is guaranteed. For birders, it is one of Sri Lanka's top spots for watching birds of prey. Search for Grey-headed Fish-eagle, Black-winged Kite, Crested Hawk-eagle, Crested Serpent-eagle, White-bellied Sea-eagle, Shikra, Common Kestrel, Brown Fish-owl and Western Marsh Harrier. Also look for Malabar Pied Hornbill, Thick-billed Flowerpecker, Plum-headed Parakeet, Common Hoopoe, Sirkeer Malkoha, Blue-faced Malkoha, Little Green Bee-eater, Barred Buttonquail, Indian Silverbill and Tricoloured Munia. Migrants you might spot include Black-capped Kingfisher, Blyth's Pipit and Orange-headed Thrush.

Day 6 Early morning, one last safari at Uda Walawe National Park for dry-zone specialities. After breakfast leave for Yala for three nights. Afternoon birding at Palatupana Salt Pans for shorebirds. Waders during the migrant season include Ruddy Turnstone, Little-ringed, Kentish, Lesser Sand, Greater Sand, Grey and Golden Plovers, Little Stint, Curlew, Common and Marsh Sandpipers and Common Redshank. In open patches you may see the odd-looking Indian Stone-curlew, and Great Thick-knee and Ashy-crowned Finch-lark.

Day 7 Morning and afternoon game drive at Yala National Park for dry-zone birding. A day's birding in the park during the northern winter can yield 100 species. Look out for dry-zone specialities like Indian Stone-curlew and Great Thick-knee, Sirkeer and Blue-faced Malkohas and Malabar Pied Hornbill. In the park's numerous waterholes Painted Stork, Lesser Whistling-duck, Black-headed Ibis, Eurasian Spoonbill, Great, Median and Little Egrets and the rare Black-necked Stork may be seen. Keep a look-out for endemics such as Ceylon Swallow, Ceylon Woodshrike and Ceylon Junglefowl. Yala is also excellent for watching larger animals, including Leopard, Sloth Bear, Asian Elephant, Mugger Crocodile, Sambar and Jackal.

Day 8 Early-morning game drive at nearby Bundala National Park. Among the large waterbirds you may see are Lesser Adjutant, Painted Stork, Asian Openbill, Eurasian Spoonbill, Black-headed Ibis and Woolly-necked Stork and many species of wader during the northern winter. Afternoon, one last safari at Yala National Park for birding and other wildlife.

Day 9 After breakfast, leave for Nuwara Eliya for two nights. Afternoon, visit Victoria Park for Himalayan migrants such as Kashmir Flycatcher, Pied Thrush, Indian Pitta and Indian Blue Robin. Endemics include Yellow-eared Bulbul, Ceylon White-eye and Dusky-blue Flycatcher. Also visit a nearby wetland for proposed endemic Black-throated Munia, Paddyfield Pipit, Pied Bush Chat, Pintail Snipe, Zitting Cisticola, Plain Prinia, Blyth's Reed Warbler and, if lucky, Pallas's Grasshopper Warbler.

Day 10 Morning, visit Horton Plains National Park for montane endemics and superb landscapes. Horton Plains is famous for World's End, a stunning viewpoint that is a sheer drop of 870 m (2,850 ft). In addition to cloud forests there are open grasslands on the lower slopes, which serve as feeding grounds for herbivores like Sambar. In the forest patches look for the northern race of the endemic Purple-faced Leaf Monkey (Bear Monkey) and Dusky Squirrel. Look out for montane endemics such as Yellow-eared Bulbul, Sri Lanka Bush Warbler, Ceylon Woodpigeon, Ceylon White-eye, Dusky-blue Flycatcher and possibly the scarce montane endemic Ceylon Whistling-thrush. Afterwards, visit Hakgala Botanical Gardens for habituated Purple-faced Leaf Monkeys and a chance to see montane birds.

Day 11 After breakfast, leave for two-night stay at Kithulgala. Afternoon, cross Kelani River in a dugout canoe and reach Kelani Valley Forest Reserve, a lowland tropical rainforest rich in endemic fauna and flora. Kelani Valley Forest Reserve is ideal for any lowland endemics and provides another chance to see Ceylon Hill-myna, Green-billed Coucal, Spot-winged Thrush, Ceylon Blue Magpie, Ceylon Spurfowl, Brown-capped Babbler and Red-faced Malkoha.

Day 12 Morning and afternoon endemic birding at Kelani Valley Forest Reserve.

Day 13 After breakfast, leave for Waikkal. Late afternoon, boat trip at Waikkal for waterbirds. Highlights include Black, Yellow and Cinnamon Bitterns, Little Green Heron and the migrant Black-capped Kingfisher.

Day 14 Transfer to International Airport.

Sri Lankan Birds
Some Facts About Sri Lankan Birds

Smallest	Pale-billed Flowerpecker.
Largest	Lesser Adjutant.
Mystery bird	The 'Devil Bird'. Makes a call at night that sounds like a woman being strangled. The top contender is the Spot-bellied Eagle-owl.
Largest flocks	Migrant ducks and waders form large flocks. Resident species such as Rose-ringed Parakeet and munias form large flocks that feed on ripening paddy.
Most intelligent	Crows – the primates of the bird world. They have more brain matter per body mass than other birds.
Most easily heard and least seen	Ceylon Spurfowl.
Largest, most reliably visible, most aseasonal and most-studied bird wave in the world	Sinharaja Bird Wave.
Least abundant resident	Black-necked Stork. Less than a dozen individuals may be present.
Endemic species density	Sri Lanka is on a par with New Guinea, but is surprisingly richer in the number of endemic birds per unit area than Borneo or Madagascar.

Endemic Birds

At present 34 birds are recognized as being endemic to Sri Lanka. They are identified in the text with the symbol ℮. These are listed below for convenience as they are often the target for visiting birders.

Partridges, quails and pheasants Family Phasianidae

1. Ceylon Spurfowl *Galloperdix bicalcarata*
2. Ceylon Junglefowl *Gallus lafayetii*

Pigeons and doves Family Columbidae

3. Ceylon Woodpigeon *Columba torringtonii*
4. Ceylon Green-pigeon *Treron pompadora*

Parrots Family Psittacidae

5. Ceylon Hanging-parrot *Loriculus beryllinus*
6. Layard's Parakeet *Psittacula calthropae*

Cuckoos Family Cuculidae

7. Green-billed Coucal *Centropus chlororhynchos*
8. Red-faced Malkoha *Phaenicophaeus pyrrhocephalus*

Owls Family Strigidae

9. Serendib Scops-owl *Otus thilohoffmanni*
10. Chestnut-backed Owlet *Glaucidium castanonotum*

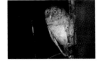

Hornbills Family Bucerotidae

11. Ceylon Grey Hornbill *Ocyceros gingalensis*

Barbets Family Capitonidae

12. Yellow-fronted Barbet *Megalaima flavifrons*
13. Ceylon Small Barbet *Megalaima rubricapillus*

Woodpeckers Family Picidae

14. Ceylon Red-backed Woodpecker *Dinopium psarodes*
15. Crimson-backed Flameback *Chrysocolaptes stricklandi*

Swallows and martins Family Hirundinidae

16. Ceylon Swallow *Hirundo hyperythra*

Cuckooshrikes Family Campephagidae

17. Ceylon Woodshrike *Tephrodornis affinis*

Bulbuls Family Pycnonotidae

18. Black-capped Bulbul *Pycnonotus melanicterus*
19. Yellow-eared Bulbul *Pycnonotus penicillatus*

Thrushes Family Turdidae

20. Spot-winged Ground-thrush *Zoothera spiloptera*
21. Ceylon Scaly Thrush *Zoothera imbricata*
22. Ceylon Whistling-thrush *Myophonus blighi*

Old World flycatchers and chats Family Muscicapidae

23. Dusky-blue Flycatcher *Eumyias sordidus*

Babblers Family Timaliidae

24. Ashy-headed Laughingthrush *Garrulax cinereifrons*
25. Brown-capped Babbler *Pellorneum fuscocapillus*
26. Ceylon Scimitar-babbler *Pomatorhinus* [*schisticeps*] *melanurus*
27. Ceylon Rufous Babbler *Turdoides rufescens*

Old World warblers Family Sylviidae

28. Sri Lanka Bush-warbler *Elaphrornis palliseri*

Flowerpeckers Family Dicaeidae

29. Legge's Flowerpecker *Dicaeum vincens*

White-eyes Family Zosteropidae
30. Ceylon White-eye *Zosterops ceylonensis*

Starlings and mynas Family Sturnidae
31. White-faced Starling *Sturnornis albofrontatus*
32. Ceylon Hill-myna *Gracula ptilogenys*

Drongos Family Dicruridae
33. Ceylon Crested Drongo *Dicrurus lophorinus*

Crows, jays, magpies and treepies Family Corvidae
34. Ceylon Blue Magpie *Urocissa ornata*

ORDERS & FAMILIES

A total of 462 bird species has been recorded in Sri Lanka at the time of writing. The table below compares the number of orders, families, genera and bird species in the world (based primarily on the IOC World Bird List v 4.2) with what is found in Sri Lanka. Bearing in mind that some orders and families are confined to certain geographical areas of the world, it can be seen that an Asian country such as Sri Lanka still holds a good proportion of birds at the level of orders and families.

	World	Sri Lanka	%
Orders	40	22	55%
Families	238	82	34%
Genera	2,273	253	11%
Species	10,680	462	4%

Just over half of the scientific orders of birds and just over a third of the scientific families have been recorded in Sri Lanka, taking into account other residents and migrants.

RESIDENTS & MIGRANTS

The number of species found only as residents or only as migrants is approximately the same. Five per cent of species have subspecies that are found as residents as well as being recorded as migrants.

	No.	%
Resident	216	47%
Migrant	220	48%
Resident & Migrant	26	5%
	462	100%

How Many Birds Can You See?

Of the 462 species recorded in Sri Lanka, 18 occur only as highly scarce migrants and 100 occur only as vagrants; a total of 118. This leaves 344 species as residents and migrants. A few of the residents and migrants are rare. If we therefore take 90 per cent of this number of 344, we get approximately 310 species. Thus, if a person who lives in Sri Lanka or is a regular visitor to the island has seen 300 species in the country, this is a good 'country list'.

For birders on a dedicated birding tour, seeing anything in excess of 200 species is good, and seeing 235 would represent a very good trip list. What is special for visiting birders to Sri Lanka is the 34 endemics, as well as the ease of viewing many species at close range.

Descriptions of Status

The status of birds described in this book derives from and has been updated from *A Checklist of the Birds of Sri Lanka* by Gehan de Silva Wijeyeratne (2007). The descriptions of status used here indicate measures of abundance (for example Common, Uncommon or Scarce), and status (for instance Resident or Migrant). The measure of abundance is subjective in the absence of quantitative data. It is based on field experience going back many years, and published information that takes into account both the geographical spread and the number of birds. Should the Square-tailed Black Bulbul, which is a highly conspicuous bird in well-visited wet zone forests like Sinharaja, be treated as a Common Resident or merely as a Resident? On an islandwide basis it would be best to describe it as an Uncommon Resident, as it is largely confined to wet zone areas, preferably with good forest stands. Similarly, Lesser Sand Plovers, though seen in large numbers, are listed as Migrants rather than Common Migrants, as they are confined to relatively small areas of suitable habitat. The abundance qualifier combined with the resident/visitor category indicates the overall status of a species.

For birds that have a threat category in relation to the IUCN Red List, this is quoted. References to the IUCN Red List are to the IUCN Red List 2007.

D. P. Wijesinghe's *Checklist of the Birds of Sri Lanka*, published in 1994, is used here as the basis for the subspecies. Any doubt on a subspecies' identity is indicated by a question mark. Some birds occur as both a resident subspecies and a migrant subspecies. This is made self-evident by the presence of two statuses.

Vagrant or Highly Scarce Migrant?

A bird that is classified as a vagrant to Sri Lanka is a bird that is not normally expected to occur as a winter visitor or passage visitor. Such birds may only be recorded a few times over a decade or even a century. Some of the vagrants have been recorded only once. A highly scarce migrant is a bird that is expected to occur in Sri Lanka regularly, but in such small numbers that in some years it may not occur at all, or its presence may fail to be recorded due to an absence of observers or limited access for observers to areas where the species is likely to be seen. In the case of some birds, such as Blue-and-white Flycatcher, establishing the status as a vagrant is easy. This is a bird that winters in Southeast Asia and there are only two records of it in South Asia; a recent record from Sri Lanka generated much interest. This is clearly a vagrant. There are other birds such as waders, and seabirds like petrels, which at the time of recording may have had only one record or just a few. However, with an increasing number of observers and higher identification skill levels, and with more people taking to sea watching or going out on pelagic trips, some of these birds may have their status revised to highly scarce migrants, scarce migrants or uncommon migrants.

At present, of the birds recorded in Sri Lanka, 18 are treated as occurring only as highly scarce migrants and 100 as vagrants. Where a bird has more than one subspecies occurring in Sri Lanka, its status is treated as that of the most abundant subspecies and is not treated as a highly scarce migrant or vagrant.

Highly Scarce Migrants

The following 18 species are highly scarce migrants:
Barau's Petrel, Lesser Frigatebird, Great Frigatebird, Christmas Frigatebird, Comb Duck, Long-legged

Buzzard, Amur Falcon, Rain Quail, 'Eastern' Baillon's Crake, Eurasian Woodcock, Jack Snipe, Great Knot, Parasitic Skua, Sandwich Tern, Lesser Noddy, Oriental Turtle-dove, Eurasian Wryneck and Citrine Wagtail.

VAGRANTS

The following 100 species are vagrants:

Cape Petrel, Bulwer's Petrel, Jouanin's Petrel, Streaked Shearwater, Sooty Shearwater, Short-tailed Shearwater, White-faced Storm-petrel, Black-bellied Storm-petrel, Red-footed Booby, Goliath Heron, Chinese Pond-heron, Eurasian Bittern, Black Stork, White Stork, Lesser Flamingo, Greylag Goose, Bar-headed Goose, Ruddy Shelduck, Gadwall, Tufted Duck, Black Baza, Egyptian Vulture, Pied Harrier, Eurasian Sparrowhawk, Tawny Eagle, Greater Spotted Eagle, Bonelli's Eagle, Lesser Kestrel, Red-headed Falcon, Eurasian Hobby, Oriental Hobby, Small Buttonquail, Brown-cheeked Water Rail, Corn Crake, Oriental Plover, Grey-headed Lapwing, Sociable Plover, Wood Snipe, Swinhoe's Snipe, Great Snipe, Spotted Redshank, Asian Dowitcher, Red Knot, Rufous-necked Stint, Sharp-tailed Sandpiper, Pectoral Sandpiper, Dunlin, Spoon-billed Sandpiper, Buff-breasted Sandpiper, South Polar Skua, Long-tailed Skua, Sooty Gull, 'Steppe' Gull, Slender-billed Gull, Black-naped Tern, White-cheeked Tern, Pale-capped Woodpigeon, Red Collared-dove, Asian Emerald Cuckoo, Great Eared-nightjar, Pacific Swift, European Roller, Greater Short-toed Lark, Dusky Crag-martin, Wire-tailed Swallow, Streak-throated Swallow, White-browed Wagtail, Tawny Pipit, Olive-backed Pipit, Red-throated Pipit, Bay-backed Shrike, Southern Grey Shrike, Eyebrowed Thrush, Yellow-rumped Flycatcher, Blue-and-white Flycatcher, Blue-throated Flycatcher, Bluethroat, Rufous-tailed Scrub-robin, Whinchat, Siberian Stonechat, Pied Wheatear, Desert Wheatear, Isabelline Wheatear, Lanceolated Warbler, Grasshopper Warbler, Booted Warbler, Dusky Warbler, Western Crowned Warbler, Indian Broad-tailed Grass-warbler, Lesser Whitethroat, 'Desert' Whitethroat, Black-headed Bunting, Red-headed Bunting, Grey-necked Bunting, Common Rosefinch, Yellow-throated Sparrow, Grey-headed Starling, Daurian Starling, Black-naped Oriole and Asian Fairy-bluebird.

ABBREVIATIONS & SYMBOLS USED

Where appropriate, captions to photographs indicate plumage, sex and/or age. In some species the plumages for adult males and females differ. In all species, juveniles are different from adults. However, from the viewpoint of identification for birders, this is usually only important in the case of birds such as gulls, which may take 2–4 years to attain adult plumage. Many species in northern latitudes that winter in Sri Lanka have marked differences between their summer (or breeding) and winter (non-breeding) plumages. In resident birds as well, a change can at times be seen between breeding and non-breeding plumages. However, the change is not so dramatic and is more common in waterbirds, in which the bare parts can take on bright colours.

COLOUR CODING THE TREE OF LIFE

Identifying species becomes easier when you become familiar with what type of bird it is that you are looking at. Scientists group all species into genera, which are in turn grouped into orders and then into families. The tree of life attempts to explain the inter-relationship of species. The various ways in which scientists attempt to do this is the science of taxonomy and is outside the scope of this book. Controversy and debate rage on how best to classify species, and as a result different bird books often classify them differently.

To help newcomers to birdwatching understand the relationships, the orders and families here have colour-coded headings. The common name of a bird is followed by its Latin name, which generally has two parts – the genus and the specific epithet. No two species can have the same Latin binomial, so the Latin names are relatively stable – and a species should have only one accepted Latin name at any given time. Common names, on the other hand, can vary a lot from one country to another and even within regions in a country. Some birds have a trinomial and are described as subspecies or geographical races.

Bird Topography

To get to grips with identifying birds, it is essential to learn the topographical terms for the parts and plumage tracts of a bird. A precise vocabulary of technical terms is essential for brevity, whether in written form or to be used in conversation when out in the field. For conciseness, the text often uses the terms in the diagrams below.

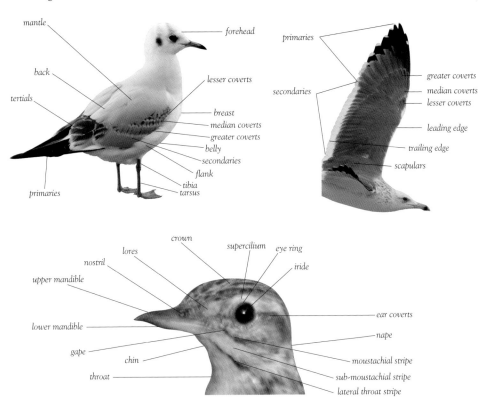

Distribution Maps

The distribution maps in this book are at best described as impressionistic. Sri Lanka does not have a network of recorders in an organized programme to log species' presence, absence or abundance into a recording grid, as is commonplace with many groups of animals and plants in European countries, for example. The only way to have a good understanding of where a bird is present is to have distribution maps on which the intensity of the colour on a recorded grid will show how many records there are. For example, in the case of a species like the Ceylon Hanging-parrot, this will show the presence of the bird in much of the wet zone in the West, but absence from the capital Colombo and its environs. It will show that it is regularly present in riverine systems even in very dry areas in the South-east and East. However, such distribution maps are only accurate if they are prepared according to a recording protocol where each recording grid is surveyed according to a predefined method and a prescribed amount of survey effort. If these colour-coded maps (also known as heat maps) are constructed from records submitted, a species may show up with a deep intensity where a number of recorders regularly submit records. However, the species may appear to be absent in areas where no observers are submitting records, even though it may be abundant in those areas.

In preparing the maps I have supplemented my own personal observations as well as having consulted the literature including other field guides which have been published in the last decade or so that included distribution maps. I will repeat the caveat that the maps are an idiosyncratic view not based on a statistically robust quantitative method and are at best indicative.

Green is used for resident birds. Blue is used for migrant birds. Solid blue indicates areas regularly occupied by a migrant. I have, in a few rare instances, used solid blue for vagrants where I have subjectively assessed the paucity of records as being due to a lack of skilled observers. I have used striping to indicate a rarity and striped blue lines indicate a rarity that might possibly turn up in the area shown. The repeat observations of rarities in specific locations are likely to be both due to the habitat being right for a species as well as being a location that is watched by birders (especially on guided tours) with advanced skills and optics.

Migrant   *Resident*   *Vagrant*

The maps in this book are colour shaded very loosely. So, for example, Ceylon Hanging-parrot and Ceylon Junglefowl may appear to be recorded in Colombo. I have not indicated that the hanging-parrot can occur in the dry zone in riverine habitat. The Ceylon Junglefowl is not found in and around cities like Colombo, not even as a bird passing through. The distribution is very high level and largely indicates the broad area that a bird is usually found in.

Sri Lanka is a small island and migrants in particular may be recorded in transit in a number of areas. For example, many wader species are present often just for a few days as flocks travel to and from suitable coastal habitats in the South of the country. Thus, records of a migrant bird being present do not signify that it takes up residence for long. Therefore, the distribution maps shown here need to be treated with caution.

Glossary

alula wing	The alula or bastard wing corresponds to the free-moving 'thumb' of a bird. It is on the leading edge of the wing at the wrist where the division is between the primary and secondary feathers.
axillaries	Feathers in 'armpit' of a bird, that is, at base of underwing.
axillary spurs	Projection of a different colour from axillaries ('armpits') into wing. A good example is the white axillary spurs in male Lesser Frigatebird.
gular	Refers to throat. See also gular stripe.
gular stripe	Vertical stripe on throat (or gular area), as seen in some birds of prey.
hepatic phase	A red or rufous colour phase in some birds, such as cuckoos and frogmouths.
heterodactyl	Describes characteristic in which a bird's third and fourth toes face forwards, and the first and second toes face backwards (only occurs in trogons).
kleptoparasitism	Usually refers to one bird stealing food from another bird species. May also refer to stealing nest material.
onomatopoeic	Term used where name of a bird is derived from its call, for example the warbler Chiffchaff.
primary projection	Extension of primaries beyond secondaries when wings are closed.
remiges	Primary and secondary flight feathers.
trousers	Feathering over feet (usually in reference to birds of prey).
zygodactyl	Describes characteristic in which a bird's second and third toes face forwards, and the first and fourth toes face backwards (as, for example, in woodpeckers and cuckoos).

PODICIPEDIFORMES
This order has only one family worldwide, the grebes.

GREBES Family *Podicipedidae*
Grebes are aquatic birds with feet at the rear of the body, which helps with propulsion in the water. They are ungainly on land.

LITTLE GREBE *Tachybaptus ruficollis capensis*

Size: 27cm

Habitat: Lowland freshwater lakes. Common bird in aquatic habitats, but easily overlooked because of its small size and discreet habits. At times, teams of Little Grebes gather to form rafts that may hold more than a hundred birds. The lakes in Annaiwilundawa are a good site for this phenomenon.

Distribution: Widespread in lowlands. Scarce in hills, though may be seen on Lake Gregory in Nuwara Eliya in highlands.

Voice: Utters a whinnying call that is heard most frequently during the breeding season.

Status: Resident.

Juvenile

In breeding plumage, neck and cheeks turn deep chestnut. Crown and upper half of face turn black. Gape develops a pale patch that can look luminescent green at close range. In non-breeding plumage, colours are duller. Sexes are similar.

Breeding

PROCELLARIIFORMES

Tube-shaped nostrils on top of the bill are the unifying characteristic of this family. Of the four families, the albatrosses have two tubes on either side, but in the other three families these are fused into a single tube.

PETRELS AND SHEARWATERS Family
Procellariidae

These are medium-sized seabirds. Most petrels are pale in colour, while shearwaters are dark. The shearwaters adroitly skim the waves, dipping in and out of wave troughs. Many of the petrels and shearwaters are vagrants, and their distribution recorded as being from the west coast is potentially a reflection of the fact that most of the resident birdwatchers in Sri Lanka are found here. When the seas are calm for whale watching, many species that are present around the time of the south-west monsoon are absent, and are not seen from the whale-watching boats that sail out from Mirissa in the South.

CAPE PETREL *Daption capense*

Size: 39cm

Habitat: Pelagic.

Distribution: A single record from Gulf of Mannar.

Status: Vagrant.

White underparts, and black head and tail. Upperparts marked strongly in black-and-white pattern. Back and centres of wings white with black spots. Edges of wings black. Unlikely to be mistaken for any other petrel.

Seabird Migration Watching

Along the west coast a mass migration of Brown-winged (Bridled) Terns, together with other species of seafaring birds, takes place, with a peak in August–September. This was first discovered and written about by the late Thilo Hoffmann and subsequently also by Arnoud Van der Bergh. Since then a detailed picture has been built due to the perseverance of local ornithologist Rex de Silva, who has studied their migration for more than two decades. Use your judgement and avoid using optics close to a sensitive site like the Colombo Harbour.

You need to be sea watching when the dawn light breaks, as the birds move beyond the horizon and out of sight an hour or so after daybreak. Bad weather brings them right in, almost to the beach. Seabirds to be seen include Bridled Tern, Brown Noddy, Pomarine Skua, Wedge-tailed Shearwater and Wilson's Storm-petrel.

On the seabird study watches arranged by Rex de Silva in and near the west coast, using telescopes from the shore, storm-petrels have been seen in and near Colombo in August–September. Storm-petrels are seen up to November. People whale watching in November (when conditions have permitted), have seen Wilson's Storm-petrels. However, when the whale watching begins in earnest from December (in the South when the seas are calmer), the storm-petrels all seem to have gone. This could be due to the flight path of storm-petrels not going through the whale-watching area, or there could have been a shift over time. More study is required on the west and south coasts to get a better understanding of the movement of the storm-petrels.

BARAU'S PETREL *Pterodroma baraui*

Size: 38cm

Habitat: Pelagic.

Distribution: Western seaboard.

Status: Highly Scarce Migrant.

Best identified by M-shaped dark formation on grey upperwing. Underwing white with M-shape discontinuous as it is interrupted by white belly. White forehead contrasts with blackish hindcrown and nape, forming a 'half-hood'. Dark tail and rump contrast with grey back.

BULWER'S PETREL *Bulweria bulwerii*

Size: 27cm

Habitat: Pelagic.

Distribution: Western seaboard.

Status: Vagrant.

This and Jouanin's Petrel (see p. 37) are small, dark petrels with long tails. Bulwer's has a more obviously paler patch on the upperwing formed by the greater coverts. Its bill is also more slender. Bulwer's has a weaker flight with rapid wingbeats and twisting glides.

JOUANIN'S PETREL *Bulweria fallax*

Size: 31cm

Habitat: Pelagic.

Distribution: Western seaboard.

Status: Vagrant.

Small, dark petrel with long tail and possible confusion with Bulwer's Petrel (see p. 36). Does not usually have pale panels on upperwings shown by Bulwer's, though in worn plumage greater coverts may be paler than rest of wing. Flight more robust, with shearing and long glides.

STREAKED SHEARWATER
Calonectris leucomelas

Size: 48cm

Habitat: Pelagic.

Distribution: A few records from western seaboard and Gulf of Mannar.

Status: Vagrant.

Key field character is white face. Head off-white and may have some mottling. Belly to vent white. Underwing has a dark trailing edge; otherwise underwing is mainly white with dark primaries and underwing primary coverts showing dark streaks. Upperparts greyish-brown.

WEDGE-TAILED SHEARWATER
Puffinus pacificus

Size: 45cm

Habitat: Pelagic.

Distribution: Arrives around mid-April before south-west monsoon. Sightings peter out by end of April, when seas become too rough for birders to go out in a boat.

Status: Migrant.

Flesh-coloured feet as in Flesh-footed Shearwater (see p. 39). More slender, all-dark bill most clearly distinguish it from Flesh-footed. Wedge-tail does not always show well. Bill pattern is the best field character. Longer-necked than Flesh-footed, but this is not always apparent.

SOOTY SHEARWATER
Puffinus griseus

Size: 43cm

Habitat: Pelagic.

Distribution: Seen on whale trips in South off Mirissa. Probably occurs as vagrant on western seaboard.

Status: Vagrant.

Dark brown on upperparts and underparts. Underwing has pale panels. In good lighting conditions these are clearly seen. With poor views, may be overlooked for Flesh-footed and Wedge-tailed Shearwaters (see p. 39 and left), which lack pale wing-panels. Legs sometimes extend beyond tail (see photographs).

FLESH-FOOTED SHEARWATER
Puffinus carneipes

Size: 47cm

Habitat: Pelagic.

Distribution: Arrives in Sri Lanka just before south-west monsoon around early to mid- April. As conditions are too rough to go out to sea on the west coast from late April, there is an absence of records after this time.

Status: Migrant.

Tail shorter and more rounded than Wedge-tailed Shearwater's (see p. 38). Legs and feet pink, but not diagnostic as shared with Wedge-tailed. Stout pink bill with dark tip is best field character. In flight in good light, primaries are supposed to show pale patch.

SHORT-TAILED SHEARWATER
Puffinus tenuirostris

Size: 42cm

Habitat: Pelagic.

Distribution: Western seaboard, with an old record from the South.

Status: Vagrant.

Overall brown shearwater. Underwing light greyish-brown, contrasting with darker leading and trailing edges. Can create a pale, panelled effect as in Sooty Shearwater (see p. 38). In Short-tailed Shearwater, legs often (but not always, see photographs) extend beyond short tail.

PERSIAN SHEARWATER
Puffinus persicus

Size: >31cm

Habitat: Pelagic.

Distribution: Seems to arrive around mid-April before south-west monsoon. Sightings peter out by end of April, when seas become too rough for birders to go out to sea. The author's photographs, taken in April 2010 off Kalpitiya, were probably the first high-quality photographs taken of this species in Sri Lankan waters, as these birds have rarely been seen. However, it is feasible that when more birdwatchers hire boats to run north–south transects between the E 79° 35' and E 79° 35' lines of longitude, many hitherto scarcely seen pelagics such as this species will be observed and photographed more frequently.

Status: Scarce Migrant.

Pale underparts and brown upperparts. Bill grey with darker tip. In April 2010, the author located a flock that held 35 Persian Shearwaters.

Storm-petrels Family *Hydrobatidae*
Petrels have a fluttering flight and some species patter their feet on the water. They are the smallest of seabirds, and are generally fast-flying birds with a keen sense of smell that they use to locate food.

WILSON'S STORM-PETREL
Oceanites oceanicus

Size: 17cm

Habitat: Pelagic waters; rarely seen close to land unless driven in by storms.

Distribution: Passage migrant seen during south-west monsoon.

Status: Migrant.

White rump distinctive and distinguishes it from Swinhoe's Storm-petrel (see p. 42), which has a dark rump. Trailing edge of wings is straight. Legs trail behind tail, which is unforked.

BLACK-BELLIED STORM-PETREL
Fregetta tropica

Size: 20cm

Habitat: Pelagic.

Distribution: Western seaboard.

Status: Vagrant.

Black head and upper body. Underwing-coverts, belly and flanks white with thick black line along centre of belly from black hood to black vent, and black tail. White uppertail-coverts contrast with dark tail.

WHITE-FACED STORM-PETREL
Pelagodroma marina

Size: 20cm

Habitat: Pelagic.

Distribution: Western seaboard.

Status: Vagrant.

White face with dark mask, white underwing-coverts and white on underparts from chin to vent. Greyish underwing flight feathers. On upperwing, flight feathers are blackish and contrast with browner forewing. Pale wing-bar formed by greater coverts.

SWINHOE'S STORM-PETREL
Oceanodroma monorhis

Size: 20cm

Habitat: Pelagic.

Distribution: Western seaboard.

Status: Scarce Migrant.

Similar in size and profile to Wilson's Storm-petrel (see p. 40), but distinguished by dark rump. Wilson's has white rump. Both species have wings angled at wrist. Pale bar on upperwing formed by greater coverts is a feature shared with Wilson's. Both species have forked tails.

Tropicbirds Family *Phaethontidae*
Graceful, tern-like seabirds that rely exclusively on marine animals like fish and squid. They nest on isolated oceanic islands and travel widely outside the breeding season. In adults the two central tail feathers are elongated. There are three species, of which two have been recorded in Sri Lanka. It is only a matter of time before Red-tailed Tropicbird *Phaethon rubricauda* is also recorded.

WHITE-TAILED TROPICBIRD
Phaethon lepturus

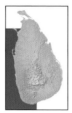

Size: 39cm w/o streamers

Habitat: Goes out to sea to fish. Roosts on trees on shoreline or on isolated small islands.

Distribution: Western and southern seaboards.

Status: Scarce Migrant.

Yellow or orange bill together with broad, V-shaped black bands on inner half of upperwings help to distinguish this species from Red-billed Tropicbird (see p. 43). Juvenile is heavily barred black and white on mantle and inner upperwings; it is similar to juvenile Red-billed, but lacks dark line to nape.

PELECANIFORMES
Five families make up this highly varied bird order.

Pelicans Family *Pelecanidae*
Pelicans are large birds with a pouch in the lower mandible used for catching fish.

RED-BILLED TROPICBIRD
Phaethon aethereus

Size: 48cm w/o streamers

Habitat: Goes out to sea to fish. Roosts on trees on shoreline or on isolated small islands.

Distribution: Western and southern seaboards.

Status: Scarce Migrant.

Both Red-billed Tropicbird and Red-tailed Tropicbird (not recorded in Sri Lanka) have red bills, and White-tailed Tropicbird (see p. 42) has yellow or orange bill. Of White-tailed and Red-billed species recorded in Sri Lanka, both have white tail-streamers, and the best way to separate them is to note that Red-billed lacks broad, V-shaped black bands on inner half of upperwings. Both species have black outer primaries, creating a black wing-tip. Juvenile similar to White-tailed but has black line to nape. In Red-billed, black on upper leading edge extends from tip to wrist. In White-tailed, black does not extend to wrist and is confined to area of primaries.

SPOT-BILLED PELICAN
Pelecanus philippensis

Size: 152cm

Habitat: Large lakes in dry lowlands. Occasionally, birds may scoop up fish in flight when water levels are low. More commonly, they are seen fishing on the water in groups.

Distribution: Bird of dry zone. Birds seen in wet zone originate from a population in the Colombo Zoo.

Status: Resident.

Duck-like gait in the water and an enormous bill. Spots on bill can be seen at close quarters. Flight feathers show only a slight contrast with white wing lining on underwing. Great White Pelican *P. onocrotalus*, with which it has been confused, has black flight feathers that contrast strongly on underwing.

MASKED BOOBY *Sula dactylatra*

Size: 80cm

Habitat: Hunts at sea. Roosts on isolated rocky islands.

Distribution: Western and southern seaboards.

Status: Scarce Migrant.

Gannets and boobies Family *Sulidae*
These birds have dagger-like bills and slim, long wings. They plunge into the water from a height to catch fish.

BROWN BOOBY
Sula leucogaster plotus

Size: 76cm

Habitat: Pelagic.

Distribution: Sri Lanka is visited by birds that breed on Indian Ocean islands. There are occasional records from seabird roosts on rocky islands on the South-western coast during the north-east monsoon. They have also been recorded off Kalpitiya. Potentially, they could turn up anywhere on the coastline.

Status: Scarce Migrant.

Adult Brown Booby has dark-edged lining underneath and brown not white on upper breast, distinguishing it from juvenile Masked Booby (see left). Masked has a white hindcollar.

Immature

Adult

Yellow bill, black mask, white body and white wings with black flight feathers. Black tail distinguishes it from white tail of Red-footed Booby (see p. 45). Juvenile similar to Brown Booby (see right) but has a white collar. Juvenile Masked Booby also has more white on leading edge of underwing. Juvenile Red-footed has entirely dark underwing.

Immature

Adult, brown morph

Juvenile

RED-FOOTED BOOBY *Sula sula*

Size: 41 cm

Habitat: Hunts at sea. Roosts on isolated rocky islands.

Distribution: Western and southern seaboards.

Status: Vagrant.

Red feet and blue bill in all four forms. 1. White form has white tail (note Masked Booby, see p. 44, has black tail), but with white wings and black flight feathers similar to those of Masked. 2. White-tailed and white-headed form has brown wings. 3. White-tailed form is brown with white tail. 4. Brown form is all brown. Note that Brown Booby (see p. 44) has white underparts (except breast to throat), and extensive white on underwing.

Cormorants and shags Family
Phalacrocoracidae
These long-necked, black-plumaged birds hunt underwater for fish. They are often seen drying out their wings.

LITTLE CORMORANT *Phalacrocorax niger*

Size: 51 cm

Habitat: Freshwater lakes and rivers in lowlands.

Distribution: Spread throughout Sri Lanka but numbers are greatest in dry lowlands

Status: Common Resident.

Bill is best field characteristic to separate it from Indian Shag (see p. 46). Bill is short and stubby in this species, slender and long in Indian Shag. Juveniles are brown with white throats. Underparts turn scaly before assuming adult plumage.

INDIAN SHAG
Phalacrocorax fuscicollis

Size: 63cm

Habitat: Freshwater lakes and rivers in lowlands. Prefers lakes to rivers and streams. At times, flocks of hundreds can gather to hunt fish.

Distribution: Spread throughout Sri Lanka but numbers are greatest in dry lowlands.

Status: Common Resident. Not as abundant as Little Cormorant (see p. 45).

Relatively long and slender bill. Bluish eyes. Juveniles are brown with white throats. Develops a little white ear-tuft in breeding plumage. At close quarters, also shows greenish irides.

Breeding

GREAT CORMORANT
Phalacrocorax carbo sinensis

Acquiring breeding plumage

Breeding

Size: 80cm

Habitat: Lakes.

Distribution: North Central and Eastern Provinces.

Status: Scarce Resident.

Breeding adults have white flank-patches. In addition to pale area around bill, feathers on crown and face surrounding bare skin around bill, turn white. Thick crescent of black behind eye remains, providing a contrast. Looks white-headed at a distance. White edging continues on back of head and along neck. Patch of orange-coloured bare facial skin present below eye; blue irides.

Darters Family *Anhingidae*
Darters are similar to cormorants, but have much longer necks and long, pointed, dagger-like bills.

ORIENTAL DARTER *Anhinga melanogaster*

Size: 90cm

Habitat: Lowland lakes.

Distribution: Mainly in dry zone.

Status: Uncommon Resident. Relatively scarce bird.

Also known as the snake bird because of its snake-like appearance when it is submerged with just the long head and bill sticking out of the water. Looks superficially similar to cormorants.

Frigatebirds Family *Fregatidae*
These birds with thin, long wings breed on oceanic islands. They occur in Sri Lanka on passage or as vagrants. Three species have been recorded so far in Sri Lanka. Adults have dark upperparts and males have a red throat-patch that they inflate during courtship displays. Identification of immature plumages can be quite tricky, and was probably not possible with certainty until the publication of a guide to the identification of these three species by David James in the June 2004 issue of *Birding Asia*.

Male

Female

LESSER FRIGATEBIRD *Fregata ariel*

Size: 76cm

Habitat: Pelagic.

Distribution: Can occur anywhere along coast. Most records are from the west coast, which may reflect the presence of observers.

Status: Highly Scarce Migrant.

Adult male all dark except for thin white axillary spurs. Female has white belly and thick white axillary spurs. Eye-ring red. Juvenile has orange-yellow head, white belly and ill-defined white axillary spurs.

GREAT FRIGATEBIRD
Fregata minor

Size: 93cm

Habitat: Pelagic.

Distribution: Can occur anywhere along coast. Most records are from the west coast, which may reflect the presence of observers.

Status: Highly Scarce Migrant.

Adult male all dark. Adult female has white on chin, throat and breast. Juveniles have white extending from whole head to belly.

Female

CHRISTMAS FRIGATEBIRD
Fregata andrewsi

Male

Female

Size: 95cm

Habitat: Pelagic.

Distribution: Bad weather may bring it close to shore during south-west monsoon (May–July).

Status: Highly Scarce Migrant.

Adult males are easy to identify as they are all dark underneath except for white belly-patch near vent. Female has white belly with a pair of broad white axillary spurs extending out to wings, and another pair of white strips extending out to sides of head (in a star shape). Immature females also have this star shape, but contrast is less. Careful examination of photographs is needed to distinguish immature Christmas Frigatebirds from immature Lesser Frigatebirds (see p. 47).

Identifiying the Frigatebirds

	Christmas	Great	Lesser
Adult male	All dark, except for white ventral patch	All dark, no pale patches on underparts	All dark, except for pair of thin white axillary spurs
Adult female	Dark, except for white belly and white 'star'	Dark, except for broad, white area from chin and throat to breast; area above eye is dark	Pair of thick white axillary spurs and white on belly; red eye-ring
Juveniles Very variable; this is only an indicative guide	Juvenile females dark, except for yellow on head bordered with white, white belly with thick, white axillary spurs; juvenile males have white from belly to vent, dark heads, and lack white axillary spurs Juveniles have dark breast band against yellow hood	Dark, except for white head and belly No yellow hood. No. dark breast band	Dark, except for white on belly and ill-defined axillary spurs No dark breast band against yellow hood

Great Frigatebird, female

Christmas Frigatebird, female

Lesser Frigatebird, female

CICONIIFORMES
This waterbird order comprises six families and
several subfamilies within them. Three families
occur in Sri Lanka.

Herons and egrets Family *Ardeidae*
These generally long-billed, long-legged birds hunt
for vertebrate and invertebrate prey at the water's
edge.

LITTLE EGRET *Egretta garzetta garzetta*

Size: 63cm

Habitat: Widespread, mainly in
lowlands. Almost any waterbody
from lagoons and rivers to canals.
Found in fresh and brackish water
habitats. Also gathers at freshly
ploughed paddy fields.

Distribution: Mainly in lowlands, but
ascends to mid-hills. Most common in dry lowlands.

Voice: As in all egrets, vocabulary limited to a few
guttural calls.

Status: Common Resident.

Bill always black and 'pencil thin'. Yellow on feet
always separates it from other egrets. At peak of
breeding, feet and lores turn flesh colour. During
breeding season develops 'aigrette' plumes
(feathers on the back of the head).

WESTERN REEF-HERON
Egretta gularis schistacea

Size: 63cm

Habitat: Coastlines and lagoons.

Distribution: Coastal wetlands, but
rare in southern half of Sri Lanka.
Most records are from Mannar
and the strip extending north from
Kalpitiya.

Status: Scarce Migrant.

Occurs in dark and white morphs with
intermediate forms as well. Grey morph grey
all over with white chin (note that Little Egret,
see left) can also have grey morph). Best field
characteristic is the bill, which is heavier than that
of Little Egret. Yellow on feet of the reef-heron
usually extends up to tibia, sometimes as far up
as hock joint; but note that in some Little Egrets
yellow can extend up the feet beyond the 'yellow
slippers'.

GREAT EGRET *Egretta alba*

Size: 90cm

Habitat: Large waterbodies and marshes in lowlands.

Distribution: Common in lowlands, and ascends to mid-hills.

Voice: In flight, often utters a low croak. Sometimes this is uttered as a protest and at times it is a contact call.

Status: Common Resident.

Separated from Intermediate Egret (see right) by gape-line that extends behind eye. Largest of the egrets. Yellow bill turns black in breeding plumage. At peak of breeding season, lores turn blue and tibia turns crimson.

Breeding

Non-breeding

Non-breeding

Breeding

INTERMEDIATE EGRET
Egretta intermedia intermedia

Size: 80cm

Habitat: Found mainly in freshwater bodies, though may occasionally fish in brackish water.

Distribution: Mainly in dry zone.

Status: Common Resident.

In breeding plumage, yellow bill turns black with a dusky tip. Separated from Little Egret (see p. 50) by heavier bill and black feet. Smaller than Great Egret (see left), and larger than Little Egret.

GREY HERON *Ardea cinerea cinerea*

Size: 96cm

Habitat: Lowland lakes and marshes. Comfortable in stretches of open water. Feeds on a variety of aquatic animals.

Distribution: Dry zone.

Voice: Harsh, loud *kraa* calls usually in flight.

Status: Resident.

GOLIATH HERON *Ardea goliath*

Size: 147cm

Habitat: Large freshwater bodies.

Distribution: May turn up anywhere in lowlands.

Voice: Harsh, guttural croaks, typically in flight when disturbed.

Status: Vagrant.

Around the size of Purple Heron (see p. 53). Has grey wings and a white neck, and head with a black side-stripe on crown. For a brief period during breeding season, legs turn crimson, bill becomes orange and it develops plumes.

Heavy grey bill and grey legs are diagnostic for separating it from superficially similar but smaller Purple Heron. In flight, note plain grey upperwings. Purple Heron has darker contrasting flight feathers.

PURPLE HERON *Ardea purpurea manilensis*

Size: 86cm

Habitat: Lowland lakes and marshes. Equally at home in swampy habitats in dry and wet zones. Prefers areas fringed with reed beds.

Distribution: Mainly in wet zone lowlands and less frequently in dry lowlands, where it feeds on a variety of aquatic animals.

Voice: Guttural *kreek* in flight; rolling series of guttural calls when interacting.

Status: Resident.

Rufous on neck with body and wings greyer. Distinguished from larger vagrant Goliath Heron (see p. 52), which has heavy grey bill. Purple Heron's bill includes yellow, which is brighter in breeding plumage.

Breeding

EASTERN CATTLE EGRET
Bubulcus coromandus

Size: 51cm

Habitat: Mainly lowland paddy fields and grassy pastures. Increasingly seen at refuse dumps.

Distribution: Found throughout Sri Lanka. In the 2000s, began to spread to highlands. Ambewala Cattle Farm near Horton Plains may be a factor that has helped this bird to become established around Horton Plains National Park. Climate change could be another factor.

Status: Common Resident.

Heavy jowl (puffy throat) and stocky build distinguish it from other egrets. Golden-buff wash on head, neck, face and breast in breeding plumage. Legs black in breeding and non-breeding birds.

INDIAN POND HERON
Ardeola grayii grayii

Size: 46cm

Habitat: Marshes, paddy fields and even cricket grounds in lowlands.

Distribution: Widespread throughout Sri Lanka up to highlands.

Voice: Usually silent; a protesting *kreek* when flushed.

Status: Common Resident. Possibly the most abundant wetland bird.

Streaked brown upperparts make the bird invisible in grassy environments. When it suddenly takes flight, its clean white wings are visible. During breeding season, fleshy parts change colour. Lores turn bluish and legs, especially around hock joint, turn crimson. Mantle turns purple. Black-tipped yellow bill in breeding and non-breeding plumages. Vagrant Chinese Pond Heron (see right) has grey back, but in non-breeding birds plumage is hard to separate.

Breeding

Acquiring breeding

CHINESE POND HERON
Ardeola bacchus

Size: 52cm

Habitat: Variety of freshwater habitats from rice fields to wet meadows and ponds.

Distribution: Recorded in lowlands.

Voice: Indignant, croaking call.

Status: Vagrant.

In breeding plumage, head and neck are deep chestnut with back dark grey to black. In Sri Lanka, migrant birds will most likely be in non-breeding plumage. Non-breeding Chinese and Indian Pond Herons (see left) are hard to distinguish in the field.

STRIATED HERON Butorides striata javanicus

Size: 44cm

Habitat: Brackish swamps and mangroves in lowland wetlands.

Distribution: Throughout coastal lowlands. Seems to be most common in coastal mangroves from Kalpitiya to Mannar.

Voice: Usually silent; flies away with staccato alarm call when disturbed.

Status: Uncommon Resident.

Dark crown contrasts with grey upperparts. Wing feathers are pale edged, giving a scalloped effect. Juveniles are brown with white spots on tips of coverts forming bars and with streaks underneath.

BLACK-CROWNED NIGHT-HERON
Nycticorax nycticorax nycticorax

Size: 58cm

Habitat: Thickets in low-lying lakes. Not confined to freshwater habitats and can be seen in brackish habitats as well, though freshwater lakes and rivers seem to be the preferred habitat.

Distribution: Found throughout lowlands to mid-hills. Presence may be overlooked on account of its nocturnal habits.

Voice: Often utters a harsh, guttural *kwok* call that betrays its presence when it is flying overhead.

Status: Uncommon Resident.

Juveniles brown spotted with white. Fleshy parts change colour at peak of breeding. Males develop fine nuchal crest and legs turn crimson.

Malayan Night-heron
Gorsachius melanolophus

Size: 51 cm

Habitat: In Sri Lanka most records are from shaded streams, though it may turn up anywhere on arrival.

Distribution: Lowlands up to mid-hills.

Voice: Variable; repeated, owl-like, throaty hoot. At times a more rasping single note.

Status: Scarce Migrant.

Juvenile

Fat body profile of a night-heron. Adult chestnut on wings and body, with black crown and nuchal crest. Dark streaks down throat to breast. In flight, chestnut wings have blackish wing-bar comprising the primaries and secondaries, with chestnut trailing edge to secondaries. Juvenile densely vermiculated with black barring on an off-white ground colour.

Yellow Bittern *Ixobrychus sinensis*

Size: 38cm

Habitat: Swamps and marshes in lowlands.

Distribution: Uncommon breeding resident in lowlands. Migrant birds spread all over Sri Lanka up to mid-hills. Commonly seen during winter.

Status: Uncommon Resident and Migrant. The most common of the four bitterns found in Sri Lanka, with a resident race as well as a migrant race.

Juvenile streakier than adults. Contrasting upperwing that shows clearly in flight. Similar Chestnut Bittern (see p. 57) lacks black flight feathers.

CHESTNUT BITTERN
Ixobrychus cinnamomeus

Size: 38cm

Habitat: Freshwater marshes and lakes with dense aquatic vegetation.

Distribution: Scarce breeding resident in lowlands; even more rare up to mid-hills.

Status: Uncommon Resident.

Uniform chestnut upperwing distinguishes this species from Yellow Bittern (see p. 56), which has black flight feathers.

Juvenile

Juvenile

BLACK BITTERN *Dupetor flavicollis flavicollis*

Size: 58cm

Habitat: Swamps and marshes in lowlands. Occasionally in forested streams.

Distribution: Rare breeding resident in lowlands. Migrant birds spread all over Sri Lanka up to mid-hills. Talangama Wetland is a reliable site for birders.

Voice: Sounds like a frog barking.

Status: Uncommon Resident and Migrant.

Black overall with thick yellow stripe on sides of neck, and thinner pale stripes on front of body running from chin to breast. Female duller than male; brownish-black. Juvenile like paler female with pale edges to wing feathers.

EURASIAN BITTERN *Botaurus stellaris*

Size: 71cm

Habitat: Wetlands with extensive reed beds where it can conceal itself.

Distribution: Most likely to turn up in lowlands.

Voice: Sometimes a barking call; famous booming call has a slight metallic resonance.

Status: Vagrant.

Chestnut ground colour heavily patterned with feathery stripes formed of arrowheads. Black crown. Confusion with juvenile Malayan Night-heron (see p. 56) is unlikely, as patterning is not so dense and body shape of Eurasian Bittern is the classic slim, bittern shape, and not the chunky night-heron shape.

Storks Family *Ciconiidae*
Storks are long-legged, long-billed large birds with thin, long wings used for gliding on thermals. They eat small animals such as frogs and lizards, and also hunt in the water for fish and aquatic invertebrates. Large storks such as the Painted Stork will even eat baby crocodiles.

PAINTED STORK
Mycteria leucocephala

Size: 93cm

Habitat: Mainly edges of open waterbodies. Wades into centres of shallow ponds. Has been known to eat baby crocodiles,

Distribution: Dry lowlands. Birds in wet zone originate from a colony in the Colombo Zoo.

Status: Resident.

Yellow bill, orange face and black-and-white body. Pink flush on rear end of body. Bill slightly curved down at tip.

ASIAN OPENBILL *Anastomus oscitans*

Size: 76cm

Habitat: Marshes, paddy fields, edges of lakes and ponds.

Distribution: Mainly in lowlands, but ascends to mid-hills.

Status: Resident.

Black-and-white plumage; superficially similar to vagrant White Stork (see p. 60) in body plumage pattern, but bill is thick with gap between mandibles; the 'open bill'. This is only clear at close range. Bill colour is a dirty white.

Juvenile

Adult left, juvenile right

BLACK STORK *Ciconia nigra*

Size: 98cm

Habitat: Favours shallow water habitats.

Distribution: A few records from dry lowlands. Records scattered from North-east, north-central region, to South-east. Records have been from thinly populated areas (mainly national parks) with large areas of grassland, scrub forest and water.

Voice: Yelping call with wheezy notes.

Status: Vagrant.

Red legs, bill and eye-patch. White underparts. Black head, neck and upperparts. Easily separated from Woolly-necked Stork (see p. 60) by its black neck and red bill (dull horn-coloured in Woolly-necked).

WOOLLY-NECKED STORK
Ciconia episcopus episcopus

Size: 83cm

Habitat: Meadows and grassland not far from water. Diet based more on terrestrial animals than aquatic animals.

Distribution: Confined to dry lowlands.

Status: Uncommon Resident.

Black with a white neck and undertail-coverts. Black plumage has a glossy sheen. Also known as White-necked Stork.

WHITE STORK *Ciconia ciconia*

Size: 105cm

Habitat: Can be found feeding in grassland, and in Europe nests in towns.

Distribution: Records in Sri Lanka are scattered in dry lowlands.

Voice: Clear, rattling call with distinct clicks.

Status: Vagrant.

Confusion possible with common Asian Openbill (see p. 59), which has the same overall black-and-white plumage. However, White Stork has a straight red bill that is diagnostic. Openbill has a curved bill that is greyish. In flight, black flight feathers contrast with white wings.

BLACK-NECKED STORK
Ephippiorhynchus asiaticus asiaticus

Size: 135cm

Habitat: Frequents lagoons and adjoining grassland. Seldom seen in fresh water.

Distribution: Largely confined to dry lowlands in South. In May 2011, Tara Wikramanayake recorded one in Trincomalee – a record of this bird in the North-east after a lapse of 50 years. In 2016, recorded in Uda Walawe N.P. A few birds present in areas of Yala and Kumana. Over the years, the number of individuals known from the South has never increased to more than half a dozen or so.

Status: Highly Scarce Resident.

Large stork with black neck. Black-and-white body. Male has brown irides, while female's irides are yellow.

Female left, male right

Male

LESSER ADJUTANT *Leptoptilos javanicus*

Size: 110cm

Habitat: Grassland, but not far from water. Often seen feeding on small, terrestrial animals like frogs and reptiles like other storks, but also known to scavenge on dead animals.

Distribution: Dry lowlands.

Status: Scarce Resident. Rare outside protected areas. Globally endangered species. Listed as Vulnerable on IUCN Red List.

Large stork with grey plumage, yellow bill and naked yellow neck.

> **Ibises and spoonbills** Family
> *Threskiornithidae*
> Ibises have distinctive downcurved bills that are
> used in probing mud. Spoonbills, as their name
> suggests, have spoon-shaped bills.

GLOSSY IBIS *Plegadis falcinellus falcinellus*

Size: 60cm

Habitat: Freshwater wetlands and mangrove areas.

Distribution: Can turn up anywhere in lowlands, from wetlands such as Talangama close to the capital Colombo, to wetlands in dry lowlands such as Kalametiya. Small flocks of up to half a dozen have been recorded. They seem to move a lot.

Status: Scarce Migrant.

The only all-dark species of ibis in Sri Lanka.

BLACK-HEADED IBIS
Threskiornis melanocephalus

Size: 69cm

Habitat: Marshes and wet fields in lowlands. Often feeds in paddy fields when they have just been ploughed or have lain fallow. Prey composition changes from soil-dwelling invertebrates to small animals and invertebrates that are on dry soil. Interestingly, G. M. Henry describes it as largely nocturnal, but I have never thought of it as such.

Distribution: Widespread in lowlands ascending to mid-hills.

Status: Resident.

Black head and neck, and white body. Downcurved bill.

EURASIAN SPOONBILL
Platalea leucorodia leucorodia

Size: 82cm

Habitat: Lagoons and freshwater bodies in dry lowlands. Eats a wide range of animals, including amphibians and insects.

Distribution: Wide distribution: found across Europe, Africa and Asia.

Status: Uncommon Resident.

Buffy-yellow breast-band in breeding plumage. During breeding season, develops crest of feathers on back of head, and its dark bill is tipped with yellow.

PHOENICOPTERIFORMES
This order has only one family worldwide – the five species of flamingo.

Flamingos Family *Phoenicopteridae*
Flamingos are distinctive pink birds with downcurved bills and long necks. They feed by holding the head upside-down, and form large flocks.

GREATER FLAMINGO *Phoenicopterus roseus*

Size: 110cm

Habitat: Salt pans and lagoons in coastal areas. A filter feeder, feeding on tiny aquatic animals and micro-organisms.

Distribution: At present mainly around Hambantota, Mannar Causeway and Northern Peninsula. There are no confirmed records of breeding in Sri Lanka, though the birds have constructed nest mounds and abandoned them. Their movements between India and Sri Lanka are not regular, and there can be a gap of a few years when they are absent. Migrants to Sri Lanka are believed to breed in the Rann of Kutch.

Voice: A flock keeps up a continuous babble of geese-like honking noises.

Status: Migrant.

Looks quite plain when the scarlet-and-black wings are tucked beneath the pale mantle.

Lesser Flamingo *Phoeniconaias minor*

Size: 90cm

Habitat: Estuaries, lagoons and salt pans, usually in company of Greater.

Distribution: Most likely to occur in Northern Peninsula, Mannar or South-east, in the company of Greater.

Voice: Honking calls, slight metallic tonality and different from that of Greater.

Status: Vagrant.

Best separated from Greater Flamingo (see p. 63) by its deep crimson bill with a black tip. The irides are red (pale in Greater). This gives the Lesser Flamingo an overall darker-headed look. Pink on Lesser is also usually richer and more vivid. Greater often looks more white than pink. Lesser is shorter than Greater. Note that juvenile Greater can be mistaken for shorter Lesser.

ANSERIFORMES
This order has two families, the Anhimidae, which are found in South America and the Anatidae, found worldwide.

Swans, geese and ducks Family *Anatidae*
Swans and ducks are familiar. The family consists of a varied mix of duck-like and swan-like waterbirds whose genetic relationships to each other are not entirely clear.

Fulvous Whistling-duck
Dendrocygna bicolour

Size: 51cm

Habitat: Aquatic habitats from lakes to marshes.

Distribution: Records have been from wet lowlands, with breeding recorded in dry lowlands in North-west of Sri Lanka.

Voice: High-pitched, double-noted, nasal call, very different from that of Lesser.

Status: Vagrant and Rare Breeding Resident. Considered a vagrant to Sri Lanka until 2015, when Tara Wikramanayake found it breeding in the Annaiwilundawa Wetland.

Separated from Lesser Whistling-duck (see p. 65) by dark line down hindneck and prominent creamy-white flank-plumes. In flight, wings are all dark. In Lesser, forewing is chestnut. Slightly bigger than Lesser, but this is not readily apparent in the field.

LESSER WHISTLING-DUCK
Dendrocygna javanica

Size: 42cm

Habitat: Marshes and lakes in lowlands. Bird of fresh water that does not use brackish waterbodies. Feeds on aquatic vegetation.

Distribution: Common throughout lowlands, ascending to mid-hills.

Voice: Also known as Whistling Teal on account of its whistling calls, usually uttered in flight. Heard quite often in flight at night.

Status: Resident.

Confusion is possible with Fulvous Whistling-duck (see p. 64). The latter has white-streaked flanks, off-white uppertail-coverts and black line on upper side of neck. Lesser has chestnut wing-coverts that contrast with dark flight feathers. Fulvous has uniformly dark upperwings.

GREYLAG GOOSE *Anser anser*

Size: 81cm

Habitat: Wet meadows and grassland. Sometimes seen swimming in streams in small flocks.

Distribution: Likely to occur in lowlands.

Voice: Fast, repeated *ahh ahh* and also longer honking calls. Chattering calls when flocking.

Status: Vagrant.

Pink feet and legs. Largely greyish-brown except for dark flight feathers and white vent. Neck streaked with dark creases. Confusion unlikely with other waterfowl in Sri Lanka. Large size.

BAR-HEADED GOOSE *Anser indicus*

Size: 75cm

Habitat: In its breeding grounds, found in high-altitude lakes. Winters by lakes and rivers.

Distribution: In Sri Lanka sight records are from Mannar, with records of grey geese from Northern Peninsula believed to relate to this, or Greylag Goose (see p. 65), or both.

Voice: Nasal honking call.

Status: Vagrant.

Two black bars on white head are distinctive. White stripe continues down grey neck from white head. Overall a pale grey bird, almost looking white. Wings also grey, with dark inner primaries and secondaries.

Female

Female

RUDDY SHELDUCK *Tadorna ferruginea*

Size: 66cm

Habitat: Like other *Tadorna* species, favours estuarine habitats.

Distribution: Vagrant to estuaries.

Voice: A few varied calls, generally quivering or indignant-sounding honks.

Status: Vagrant.

Large duck with overall ruddy colour; head slightly paler. Forewing is white, contrasting strongly with black flight feathers. Underwing is black and white with white wing lining and dark flight feathers. In India found in clean rivers. Male has black neck collar.

COMB DUCK *Sarkidiornis melanotos*

Size: 76cm

Habitat: Freshwater bodies in dry lowlands.

Distribution: Several records from South-west and records from Mannar and Northern Peninsula.

Voice: High-pitched, nasal call. Also a gutteral *kraa*.

Status: Highly Scarce Migrant.

White duck with glossy black wings and black-and-white mottled cheeks and neck. Dark spur from shoulder to belly. Black crown. Both sexes have lead-coloured bills. Male develops a comb on upper mandible. Juvenile is brown.

Female

Male

Male

Female

COTTON TEAL *Nettapus coromandelianus*

Size: 33cm

Habitat: Lily-covered lakes in lowlands.

Distribution: Seems to be most common in dry lowlands.

Status: Uncommon Resident.

Diminutive duck. Male has a green back, and white neck and head with a dark crown and black collar. Female is brown and white with dark line through eye. Male has prominent white wing-bar, lacking in female. Females always have one or more males associating with them.

GADWALL *Anas strepera*

Size: 51cm

Habitat: Freshwater lakes and ponds.

Distribution: Rare visitor to wetlands.

Voice: A bleating *quack* interspersed with whistles.

Status: Scarce Migrant.

Male is greyish, and female is browner. Male has a grey bill and female an orange-sided bill, which helps to distinguish it from other female ducks. Both sexes have white wing-bar on inner secondaries, which shows as a white flash when on the water. Male's tail is fringed with a black vent and uppertail-coverts. Grey on male is delicately vermiculated with thin, wavy lines, visible only at close range.

Male

Female

Male

Female

EURASIAN WIGEON
Anas penelope

Size: 49cm

Habitat: Winter visitor to coastal wetlands.

Distribution: Very rare in South, but flocks of several hundred are seen in northern coastal wetlands.

Voice: Males have a clear, whistled *whee oh*.

Status: Migrant.

Male has conspicuous white forewing patches. Breeding male has buffish-yellow forehead, chestnut crown and pink breast. Round-headed with small bill. In eclipse plumage, male looks like female, without buffish-yellow crown. In breeding plumage, male develops a delicate pink on the breast, which contrasts subtly with the light grey flanks and back. Horizontal white line may show on inner secondaries.

INDIAN SPOT-BILLED DUCK
Anas poecilorhyncha poecilorhyncha

Size: 61 cm

Habitat: Freshwater bodies.

Distribution: Around Mannar and Northern Peninsula. Mannar remains the best place to see this rare bird in Sri Lanka.

Status: Highly Scarce Resident and Scarce Migrant. Formerly considered a scarce winter migrant. Young were seen in 2003 in the Talladi Ponds, near the Mannar Causeway.

This suggests that it could be a breeding resident in these areas. Highly Endangered on IUCN Red List.

Large size and yellow band towards tip of grey-black bill, scalloped sides and upperparts, and heavy spotting on breast. Green speculum bordered by thin white wing-bar in front, and thick white patch formed by tertials show well in flight. Sexes are similar, with male having a larger red loral patch than female.

Male

Female

NORTHERN SHOVELER   *Anas clypeata*

Size: 51 cm

Habitat: Coastal wetlands.

Distribution: In South, around Mannar and Northern Peninsula.

Voice: Punchy, sibilant honk, rapidly repeated.

Status: Scarce Migrant.

In non-breeding plumage, male similar to female, but retains pale blue on upper wing. Male has darker belly than female. Female's flanks are buff with thick, dark brown (looks black) tips to feathers creating a scaly pattern. Feathers on upperparts are dark with pale edges creating a scaly effect. Pattern on male is different from that on female, something that does not seem to be depicted well in many illustrated field guides. Sieves water by rapidly moving bill from side to side while swimming. Often a flock of birds whirls around together. Up-ends occasionally.

NORTHERN PINTAIL
Anas acuta acuta

Size: 56–74cm

Habitat: Large flocks often occupy waterbodies, both fresh and brackish, in lowlands.

Distribution: Winter visitor in large numbers to coastal wetlands in South (for instance at Bundala), and the strip from Mannar to Northern Peninsula.

Voice: A few varied calls, with a whistled *kloop* at rest.

Status: Migrant.

Male

Male

Male

In non-breeding plumage, male is similar to female, but retains grey on upper wing. Lead-grey bill distinguishes nondescript female from other ducks. In breeding plumage, male develops a chocolate-brown hood with a thin white finger running up along neck. The 'pin tail' also develops.

GARGANEY *Anas querquedula*

Size: 41cm

Habitat: Freshwater lakes in lowlands, especially in dry lowlands. Seems equally at home in freshwater bodies and brackish water.

Distribution: Common migrant to freshwater wetlands in dry lowlands, and less common in wet lowlands. May also be seen in brackish water, but this is not its preferred habitat.

Voice: A few varied calls; one is a rapid, honking call.

Status: Migrant. Abundant winter visitor.

In non-breeding plumage, male similar to female but separated from female by bluish-grey forewing. In female, leading white border on speculum not as prominent, leaving distinct white trailing edge. In breeding plumage, male acquires a prominent white eyebrow. Rare migrant Gadwall female (see p. 68) has orange in its bill and a white patch on inner secondaries. Female Common Teal (see p. 71) has white stripe on sides of undertail-coverts, absent in Garganey. Female Common Teal is 'plain faced'.

Female

COMMON TEAL *Anas crecca crecca*

Size: 38cm

Habitat: Coastal wetlands.

Distribution: Most likely to be seen in northern wetlands around Mannar and Jaffna, where individuals may be encountered in mixed flocks of ducks. Other likely locations include Bundala in South.

Voice: High-pitched, whistled note spaced at short intervals.

Status: Scarce Migrant.

Small duck, easily overlooked in mixed duck flocks. Male has chestnut head and green eye-patch. In eclipse plumage looks like drabber female. Male has spotting on breast.

Male

Female

Female

Male left, female right

Male

TUFTED DUCK *Aythya fuligula*

Size: 43cm

Habitat: Frequents open freshwater bodies. On its breeding grounds occupies large, open lakes.

Voice: Grating *quack quack*, most often uttered in flight.

Status: Vagrant.

Male is black with white sides and underparts. Female is brown. Both sexes have yellow irides. 'Tuft' is a short nuchal crest that is not apparent most of the time.

FALCONIFORMES

This is a broad range of birds of prey, from heavy-bodied vultures to slender-built kites. Raptors (another name for them) seize their prey with their feet rather than the bill. A hook-tipped bill and talons on the toes are a characteristic of raptors.

Hawks, kites, eagles and vultures Family *Accipitridae*

These birds are distinguished from falcons by having a distinctive brow-ridge. Eight groups are loosely recognized: true hawks, buzzards, true eagles, kites and honey-buzzards, Old World vultures, fish-eagles, snake-eagles and serpent-eagles, and harriers and harrier-hawks.

JERDON'S BAZA *Aviceda jerdoni*

Size: 48cm

Habitat: Well-wooded forests.

Distribution: In Sri Lanka, confined to well-wooded areas in highlands and higher elevations in Sinharaja.

Voice: High-pitched, double-noted whistle. Not a typical raptor call.

Status: Scarce Resident.

Black mesial stripe on white throat. Female has rufous breast-band and rufous bars on white belly. Male has longitudinal black streaks on breast. Tail has thick bars. Underwing trailing edge black, and tail tipped with white in both sexes. Male's wings have more barring on flight feathers than female's. Crest usually erected when perched.

BLACK BAZA *Aviceda leuphotes*

Size: 33cm

Habitat: Both dry and wet zone forests.

Distribution: May turn up anywhere in remaining forest.

Voice: Double-noted *whee-oh*, reminiscent of a parakeet's call.

Status: Highly Scarce Migrant.

Strikingly marked baza with black head, back and wings. Contrasting breast-bands of white, edged with blackish-brown. Belly barred chestnut. Black vent. Wings with bold white patch. Upperwing flight feathers edged with chestnut. Black crest erect when perched.

ORIENTAL HONEY-BUZZARD
Pernis ptilorhynchus

Size: 65cm

Habitat: Found in areas where pockets of jungle are interspersed with open areas. As the name suggests, likes raiding honeycombs.

Distribution: Throughout Sri Lanka.

Voice: Short, high-pitched wail.

Status: Uncommon Resident and Scarce Migrant. Occurs as resident race *ruficollis*, which is supplemented by migrant race *orientalis* during northern winter.

Identified by relatively small-sized head, giving it a pigeon-headed look. A dark phase and a light phase occur. Colouration is highly variable.

BLACK-WINGED KITE
Elanus caeruleus vociferous

Size: 33cm

Habitat: Open grassland. Sometimes seen perched on overhead wires.

Distribution: Widespread; most common in dry zone and hills. It was never particularly abundant, but appears to have been declining over the years. It is possible that the use of pesticides has reduced its prey base.

Voice: One call is a nasal *hweet*. It does not sound typically raptor-like.

Status: Uncommon Resident.

Small white raptor with black shoulders. Habit of hovering like a kestrel.

BLACK KITE *Milvus migrans govinda*

Size: 61 cm

Habitat: Open areas.

Distribution: Mainly in Northern Peninsula along coast. Occasionally turns up in dry lowlands of South. Diet is mainly small mammals, birds and invertebrates, but it can be seen scavenging at refuse dumps.

Status: Scarce Resident and Migrant.

Uniformly brown. Forked tail separates it from juvenile Brahminy Kite (see right). Black-eared Kite may also occur. Currently treated as subspecies *lineatus*, but may be elevated to full species. Adult identified by pale crown and throat, and rufous suffusion on body that is more boldy streaked below; bigger than juvenile, with wider wings. All stages have dark eyes and dark ear-coverts. Juvenile has heavy whitish streaks below.

BRAHMINY KITE
Haliastur indus indus

Juvenile

Size: 48cm

Habitat: Hunts over open water, paddy fields and open areas. Seems to favour areas near water. Needs tall trees for nesting. Marked preference for fish, but will take other small animals.

Distribution: Lowlands and mid-hills.

Voice: Generally a silent bird. When gathering to roost often utters a nasal yelping call from overhead.

Status: Resident.

Adults have white head and brick-coloured wings. Juveniles are brown and may be confused with Black Kite (see left), but this has a forked tail. Juvenile Brahminy has dark underwing-coverts that contrast with underwing, and lacks dark tail-tip of juvenile Black Kite.

WHITE-BELLIED SEA-EAGLE
Haliaeetus leucogaster

Size: 68cm

Habitat: Found near waterbodies, especially in coastal areas. Mainly a bird of dry lowlands, though a few individuals occur in wet zone up to mid-hills. Feeds mainly on fish, but also takes small mammals.

Distribution: Found mainly in dry lowlands but can occur anywhere.

Voice: Rapidly repeated, agitated notes that sound as though more than one bird is calling.

Status: Uncommon Resident.

The largest of Sri Lanka's eagles. Adults have white bodies and grey wings. Juveniles are brown. In flight, holds its wings at a strong dihedral (upturned into a shallow V).

GREY-HEADED FISH-EAGLE
Ichthyophaga ichthyaetus

Size: 74cm

Habitat: Always found around open lakes with tall trees for roosting and nesting.

Distribution: Dry lowlands.

Voice: During the breeding season pairs duet, engaging in a series of melancholy and haunting calls, some of which sound like a person being strangled. In Uda Walawe National Park they have been heard calling in the night.

Status: Scarce Resident. A nationally endangered bird

Large eagle with grey head and brown wings.

EGYPTIAN VULTURE
Neophron percnopterus

Size: 64cm

Habitat: Open country, but also mountains.

Distribution: May turn up anywhere. Records from Northern Peninsula and highlands.

Status: Vagrant.

Bare-skinned vulture-face. Overall white bird with black flight feathers. Underwing pattern is the same as that in White-bellied Sea-eagle (see p. 75), with white underwing-coverts and black flight feathers. However, tail in Egyptian Vulture is wedge shaped, and head shape is different. Some reports of this species are misidentifications of White-bellied Sea-eagle.

CRESTED SERPENT-EAGLE
Spilornis cheela spilogaster

Size: 74cm

Habitat: Forests and scrub. Preys on a wide variety of small animals.

Distribution: Throughout Sri Lanka.

Voice: Double-noted call repeated, more strident and higher-pitched than similar call of Crested Hawk-eagle.

Status: Resident.

Yellow irides and cere, brown body with white spotting, dark crown and distinct nuchal crest help distinguish this species from Crested Hawk-eagle (see p. 85). Juvenile Crested Serpent-eagles that appear whitish can be distinguished from juvenile Crested Hawk-eagles by the former not having feathered legs.

WESTERN MARSH HARRIER
Circus aeruginosus

Size: 56cm

Habitat: Typically found in marshes with reed beds. Also on farmland.

Distribution: Lowlands to mid-hills.

Voice: Rapid *kik-kik-kik*.

Status: Uncommon Migrant.

Male has characteristic tricoloured pattern on upperparts. Head, mantle and upperwing-coverts are brown, wing-tips are black and rest of wings and tail are grey. Female largely brown with a paler head and mask through eye. Adult females usually have light underwing-coverts. Juveniles have dark underwing-coverts that may contrast with flight feathers. However, there is much variability, with some adult females as dark as juveniles. Soars with outer primaries open; long wings.

Male

Female

Male

PALLID HARRIER *Circus macrourus*

Size: 48cm

Habitat: Open grassland, where it hunts for rodents and birds.

Distribution: Winter visitor throughout Sri Lanka to dry lowlands.

Status: Uncommon Migrant.

Adult male easily distinguished from adult male Montagu's Harrier (see p. 78) by the latter having a black bar on base of upper secondaries. Upperparts are grey in both males. Females and juveniles are in shades of brown, and their identification is quite tricky. Pallid has tip of wing formed by primaries 2–4, prominent pale collar, penultimate band on undertail that appears as dark blob, and dark secondaries. Female underwing primaries have irregular barring and lack dark trailing edge found in female Montagu's. Axillaries are not prominently barred. Juvenile Pallid and Montagu's have buff-orange underparts and barred underwings. Juvenile Pallid often, but not always, lacks 'dark fingers' on primary wing-tips. More advanced texts should be consulted for identification of juvenile and female harriers. Male Hen Harrier *Circus cyaneus* (not recorded in Sri Lanka) is similar but has a thick black tip to primaries unlike black wedge in Pallid.

PIED HARRIER *Circus melanoleucos*

Size: 48cm

Habitat: Marshes, farmland and open country.

Distribution: Winter visitor to suitable habitats from lowlands to highlands.

Status: Vagrant.

Male

Adult male is distinctive, with black on head, wing-tips, breast and mantle on otherwise grey body (except for white leading edge of wing). Female has white uppertail-coverts and grey tail with dark barring. Upperwing flight feathers are also grey with blackish barring.

Juvenile

Male

MONTAGU'S HARRIER *Circus pygargus*

Size: 48cm

Habitat: Hunts over open grassland.

Distribution: Migrant to open areas of dry zone. Most common in Northern Peninsula. Feeds on small animals, especially rodents and birds found in open areas.

Status: Uncommon Migrant.

See Pallid Harrier (p. 77) for notes on identification. Male with grey upperwing and thin black wing-bar on secondaries is easily distinguished from male Pallid. Female's underwing has primaries with cleaner barring, and has a dark trailing edge. Female has prominent barring on axillaries. Juvenile is buff-orange underneath with 'dark fingers' formed by dark outer primary feathers. However, some juvenile Pallids also show this. Montagu's has shorter legs than Pallid.

Female

CRESTED GOSHAWK *Accipiter trivirgatus*

Size: 43cm

Habitat: Bird of tall, mature forest.

Distribution: Wide distribution from lowlands to highlands, but scarce throughout range.

Voice: Sequence of sharp, high-pitched, single-noted whistles.

Status: Uncommon Resident.

Clear black mesial line on throat. Northern Goshawk *Accipiter gentilis* recorded in India (not Sri Lanka) lacks this. Grey on face sharply demarcated from white cheeks. Tail has four broad bars. Throat streaked, breast barred. Large size compared with Besra and Shikra (see p. 80 and right) not always reliably judged in field. In flight, broad, rounded wings are pinched in at waist. Piercing yellow eyes. Juveniles have spotted tibia coverts and a large pale patch on hind area of neck, almost forming a hindcollar.

Accipiters vs Falcons

The Shikra and Sparrowhawk are accipiters that can be distinguished from the falcons by the length of the folded wing versus the length of the tail. In the accipiters the wings reach around half the length of the tail. In the falcons the wings equal or nearly reach the length of the tail-tip.

SHIKRA *Accipiter badius badius*

Size: 35cm

Habitat: Forests and wooded parts of towns and cities. Needs pockets of dense cover. Preys on small animals including birds, reptiles, mammals and amphibians.

Distribution: Found throughout Sri Lanka.

Voice: Shikra occasionally takes to soaring and uttering its loud, screaming call from the air. Call is usually a series of shrill, single *kiew kiew kiew*, or double-noted *ki-kiew ki-kiew ki-kiew* notes.

Status: Resident.

Juvenile

In similar Besra (see p. 80), white on throat is sharply demarcated from grey on sides of face. Adult Shikra lacks uppertail barring found in Besra; this is the best identification feature. Shikra also has narrower rufous barring on underparts. Faint grey gular-stripe in adult. Occasionally takes to soaring. Female larger than male. Juvenile has more distinct throat-stripe than adults, and has longitudinal streaks on underparts.

BESRA SPARROWHAWK
Accipiter virgatus

Size: 32cm

Habitat: Confined to good-quality forests. Not recorded in urban habitats, which Shikra frequents.

Distribution: Wide distribution from lowlands to highlands, but scarce throughout range.

Voice: Repeated short, single high notes. Also quick succession of high-pitched yelping notes.

Status: Scarce Resident.

Separated from similar Shikra (see p. 79) by strongly barred upper tail in adult. Also has coarser rufous barring on underparts. Grey on face is more sharply demarcated from white cheeks. Shorter primary projection than in Shikra. Distinct gular stripe; in Shikra it is an indistinct grey in adult. Call is also different. Female larger. Juvenile browner and heavily marked on underparts.

EURASIAN SPARROWHAWK
Accipiter nisus

Male

Female

Size: 34cm

Habitat: Forested habitats. In Europe can be found in cities that have tall trees, and occupies the niche taken by Shikra in Sri Lanka.

Distribution: Could potentially turn up in any wooded area.

Voice: Series of high-pitched, single notes.

Status: Vagrant.

Adult upperparts darker than in Shikra (see p. 79). More uniformly barred on underparts than in Besra (see left). Lacks strong throat-stripe of Besra. Tail prominently barred and underwing strongly barred, unlike in Shikra. Male slate-grey above with orange-red barring on underparts, and rufous cheeks and breast. Female brown above with blackish barring on underparts. Lacks rufous cheeks and breast, and has mainly white throat area with breast area barred blackish. Juvenile brown above with coarse, brownish barring on underparts, and longitudinal streaks on chin and throat.

HIMALAYAN BUZZARD Buteo burmanicus

Size: 51–57cm

Habitat: Often found where forest cover borders grassland. A medium-sized bird of prey, it will take small animals such as birds, mammals and reptiles.

Distribution: Migrant to highland areas such as Horton Plains National Park.

Status: Scarce Migrant.

Medium-sized in relation to other birds of prey found in Sri Lanka. Brown overall. Rounded head is best guide. May be confused with juvenile Brahminy Kite (see p. 74), but underwing pattern of juvenile Brahminy Kite is different. Buzzards wintering in Sri Lanka have different morphs. Advanced texts should be consulted for detailed identification.

LONG-LEGGED BUZZARD Buteo rufinus

Size: 61cm

Habitat: Occupies a variety of habitats from dry, open country to mountains.

Distribution: Records are from higher hills and highlands.

Voice: Nasal *pee yoo* mewing call.

Status: Highly Scarce Migrant.

Rufous belly and 'trousers'. Dark-phase adult has dark wing lining and white flight feathers with trailing edge in black, and black wing-tips. Dark carpal patch. Light-phase adult has brown wing lining, with rest of wing pattern similar to that of dark phase. Long-necked with heavy bill. Flight heavy and eagle-like.

BLACK EAGLE
Ictinaetus malayensis perniger

Size: 75cm

Habitat: Forests and high ridges with forest cover. Takes small animals such as birds, mammals and lizards. Giant Squirrels seem to be a preferred prey item.

Distribution: Throughout forested areas, but mainly in wet zone and especially in mid-hills and highlands.

Status: Uncommon Resident.

Medium-sized, dark eagle with yellow cere and feet.

Juvenile

GREATER SPOTTED EAGLE *Clanga clanga*

Size: 67cm

Habitat: Prefers wetland habitats, breeding near water. Can be found visiting refuse pits to scavenge for food.

Distribution: In Sri Lanka has been recorded in South-east, but could appear anywhere in lowlands.

Voice: Yapping *kyay-kyak-kyak*.

Status: Vagrant.

Dark, bulky eagle. Underparts dark brown, and flight feathers at times slightly paler. In similar Lesser Spotted Eagle *Clanga pomarina* (not recorded in Sri Lanka), flight feathers are darker than brown underparts. Upperwing-coverts show little contrast with flight feathers, unlike in Lesser Spotted in which coverts are paler. In flight, primaries lack pale patch seen in Lesser Spotted and Steppe Eagle *Aquila nipalensis* (not recorded in Sri Lanka). Adult Greater Spotted lacks pale trailing edge to wing found in adult Steppe. Juvenile has pale spots on trailing edges of coverts forming 4–5 wing-bars.

TAWNY EAGLE *Aquila rapax*

Size: 67cm

Habitat: Occupies dry, open country with scattered trees and mountains.

Distribution: Recorded in dry lowlands, but could turn up anywhere.

Voice: Barking *kiok*.

Status: Vagrant.

Adult pale with tail and flight feathers densely but indistinctly barred. Prominent yellow gape-flange extends back to below centre of eye. Baggy 'trousers'. Confusion possible with Steppe Eagle *Aquila nipalensis* (not yet recorded in Sri Lanka). In Steppe, yellow gape-line extends to rear of eye. Hand (from wrist to wing-tip) is longer in Steppe. Darker irides in Steppe.

BONELLI'S EAGLE *Hieraaetus fasciatus*

Size: 70cm

Habitat: Mix of habitats from forested hills to arid plains and rocky mountains.

Distribution: May turn up anywhere in Sri Lanka.

Voice: Nasal, repeated *yip-yip* call.

Status: Vagrant.

To avoid confusion with Himalayan Buzzard (see p. 81), note legs are feathered all the way down. Furthermore, at rest wings only reach midway to tail. In buzzards, folded wing-tips reach tail. Adult has broad black terminal tail-band, and underwing has broad, diagonal black wing-bar extending to carpal joint. Adult has white underparts streaked with black teardrops. Juvenile has buff underparts and finely barred undertail, with black terminal band absent.

BOOTED EAGLE *Hieraaetus pennatus*

Size: 52cm

Habitat: In breeding range favours deciduous and pine woodlands. In wintering range can be found in open country.

Distribution: May turn up in highlands as well as in open country in dry lowlands.

Voice: Shrill *pip-pip*.

Status: Scarce Migrant.

White 'landing lights' at base of leading edge of wing, but note that marsh harriers and honey-buzzards can also show this. Upperwing median coverts form pale patches. Pale morph has black flight feathers and white wing lining. Dark morph has brown wing lining, and dark flight feathers with a black wing-bar formed by dark trailing edge of wing lining.

RUFOUS-BELLIED EAGLE
Hieraaetus kienerii kienerii

Juvenile

Size: 57cm

Habitat: Generally seen where there are good-quality patches of forest.

Distribution: Can turn up almost anywhere in interior of Sri Lanka away from coast. Most likely to be seen in ridges that are close to large areas of forest.

Status: Uncommon Resident.

Adult is distinctive, with rufous underparts and wing lining. Chin and throat are white, and face and crown are dark, giving it a 'hooded' look. Juvenile could be confused with juvenile Crested Hawk-eagle (see p. 85). Latter has a longer tail, its wings are not as broad and its appearance is slimmer.

CRESTED HAWK-EAGLE
Spizaetus cirrhatus ceylanensis

Size: 71cm

Habitat: Forested areas. Often perches on lone trees in open areas, but needs good forest thickets.

Distribution: Throughout Sri Lanka, but most common in dry lowlands. Varied diet of small animals such as birds, mammals and lizards. Does not seem to occupy as broad a range of disturbed habitats as Crested Serpent-eagle (see p. 76).

Voice: Whistled sequence of three notes, *whew whe o.* Also longer series of 3–6 notes that ascend as *he he he he.*

Status: Resident.

Medium-sized eagle with pale underparts and a crest. Juveniles are pale and the birds get progressively darker as they age. Adults have a lot of dark streaking on the throat and breast.

LEGGE'S HAWK-EAGLE
Nisaetus kelaarti

Size: 70cm

Habitat: Forested areas in mountains and mid-hills.

Distribution: Mid-hills and highlands.

Status: Scarce Resident. Vulnerable on IUCN Red List.

Dark hood and rufous barring distinguish this species from other large eagles. Previously treated as Mountain Hawk-eagle. Now split into a species that is found in South-west India and Sri Lanka.

Osprey Family *Pandionidae*
The Osprey is the sole member of this family, and is found across the world in temperate and tropical environments. It is totally adapted to a diet of fish, which is caught by plunging into the water from the air.

OSPREY *Pandion haliaetus*

Size: 56cm

Habitat: Large freshwater lakes with tall trees providing a perch.

Distribution: Most likely to be seen on man-made lakes in dry lowlands. Reservoir in the Uda Walawe National Park is a good example of a typical site.

Voice: Shrill *keek-keek*. Also a more insistent *yip-yip*.

Status: Scarce Migrant.

Brown upperparts including upperwings, black mask on white head, white underparts and underwing flight feathers barred brown. Long wings are angled at 'wrist'. Unlikely to be confused with any other raptor.

Falcons Family *Falconidae*
Falcons lack a distinctive brow-ridge, and are generally long-winged and open-country birds.

LESSER KESTREL *Falco naumanni*

Male

Female   Male

Size: 34cm

Habitat: Rocky gorges and open country. In its breeding range will use old buildings as a nest site.

Distribution: Most records are from dry lowlands.

Voice: High-pitched *kik-kik* rattle.

Status: Vagrant.

Both in flight and at rest, male can be distinguished from female by grey panel on bases of secondaries on upperwing. Both sexes have whitish claws; Common Kestrel (see p. 87) has black claws. This is the best diagnostic for separating the females. Male has a clean grey head without a cheek-stripe as in Common. Male is also more buff on underparts, and more sparsely spotted with small spots. Smaller and slimmer than Common, but this may not be apparent in the field.

Male

COMMON KESTREL Falco tinnunculus

Size: 36cm

Habitat: Open country. Distinctive habit of hovering.

Distribution: Migrant race *tinnunculus* spreads throughout Sri Lanka on arrival.

Voice: Flight call is a rapid series of 4–5 *yip yip* notes.

Status: Migrant, Highly Scarce Migrant and Highly Scarce Resident.

Female

Three races have been recorded in Sri Lanka: *interstinctus*, a highly scarce migrant, *objurgatus*, a highly scarce resident and *tinnunculus*, a migrant. They are not distinguishable in the field. Male has a grey head and black-tipped grey tail. Female has a rufous tail barred in black. In vagrant Lesser Kestrel (see p. 86) folded wing-tips nearly reach tip of tail, and it has pale rather than black claws. Adult male Lesser is easier to identify, as it has rufous upperparts without black markings.

RED-HEADED (RED-NECKED) FALCON Falco chicquera

Size: 34cm

Habitat: Open country with trees for nesting.

Distribution: Recorded in both dry and wet lowlands, sometimes sharing a wintering range with kestrels.

Voice: Shrill, hoarse, *kee-kee*.

Status: Vagrant.

Red head, grey upperwing blackish on outer half, white chin and throat, and rest of underparts white, barred with black. Underwing pale with thin blackish barring. Tail has a black subterminal band and thin white tip. Greyish appearance makes it superficially similar to Amur Falcon (see p. 88), but this lacks red head and has rufous undertail-coverts. Male Amur also has white underwing-coverts contrasting strongly with dark flight feathers.

AMUR FALCON *Falco amurensis*

Size: 30cm

Habitat: Open country. Hunts in a manner reminiscent of a kestrel, by hovering occasionally. Hunts into dusk.

Distribution: Scarce straggler to dry, open lowlands.

Voice: Drawn out, high-pitched notes with a slight trill.

Status: Highly Scarce Migrant.

Both immatures and adults have pale red or orangeish cere, eye-ring, legs and feet. Adult male is slate-grey underneath, with reddish vent and dark flight feathers contrasting with white underwing-coverts. Female is white underneath with streaks and arrows, and has a reddish vent.

Male

Female

Juvenile

EURASIAN HOBBY *Falco subbuteo*

Size: 33cm

Habitat: Raptor of open country bordered with wooded thickets. Famous for catching dragonflies over wetlands.

Distribution: May turn up in open country in lowlands as well as highlands.

Voice: Clear, high-pitched *teew-teuw* that sounds urgent.

Status: Vagrant.

Slim falcon. Adult has red thighs and undertail-coverts contrasting with pale throat and breast area with longitudinal streaks. Upperparts slaty. Face has prominent black moustachial stripe and another black stripe behind eye. Underwing barred white with flecks on brown ground colour. Juvenile brown with light spots on edges of wing-coverts. Underparts pale, streaked with brown.

ORIENTAL HOBBY *Falco severus*

Size: 29cm

Habitat: Hunts in open country in the same fashion as Eurasian.

Distribution: Several records from highlands before the 1960s. This may reflect the distribution of British tea planters who were birdwatchers. May also turn up in lowlands.

Voice: High-pitched, drawn out *kee-kee*. Longer drawn-out than in Eurasian, and notes are less pure and a little wheezy.

Status: Vagrant.

Slightly stockier than Eurasian Hobby (see p. 88), and adult is distinguished by complete black hood and all underparts being rufous. Underwing lining also rufous. Much darker, blue-black on upperparts, compared with Eurasian. Juvenile also has rufous tinge on underparts.

SHAHEEN FALCON
Falco peregrinus peregrinator

Size: 43cm

Habitat: Sites such as Sigiriya, Bible Rock and Yapahuwa, where rocky inselbergs allow it to launch into thermals from a height, are reliable locations for it. Sometimes takes temporary residence in tall buildings in Colombo, which are like artificial vertical rock faces. Preys mainly on birds, which it captures in flight.

Distribution: In lowlands and mid-hills where suitable hunting posts in the form of cliff faces and inselbergs are found.

Voice: Rapidly repeated single note, loud and medium in pitch.

Status: Uncommon Resident. Vulnerable on IUCN Red List.

Resident subspecies of Peregrine Falcon (see p. 90). Smaller and darker than Peregrine, with blackish hood and upperparts. Unlike smaller Oriental Hobby (see left), has rufous underwing-coverts with a hint of barring. Yellow legs, yellow cere and broad yellow eye-ring.

PEREGRINE FALCON *Falco peregrinus calidus*

Size: 43cm

Habitat: Open country and prefers 'cliff habitats' like the resident race, the Shaheen.

Distribution: Migrant race may turn up anywhere on the island.

Voice: Rapidly repeated single note, loud and medium in pitch.

Status: Scarce Migrant.

The migrant race is for convenience treated here separately as the Peregrine Falcon. White underparts barred with white. Pointed wings and broad-based, short tail. Distinctive black moustachial stripe. Shaheen Falcon (see p. 89) has rufous underparts and thighs that easily separate it from migrant Peregrine. The differences in size (Peregrine is larger) may not be apparent in the field.

GALLIFORMES
This order comprises six families, with several restricted to certain parts of the world. The family found in Sri Lanka has a wide distribution.

Partridges, quails and pheasants Family *Phasianidae*
Partridges and quails are ground-dwelling birds that largely rely on cryptic camouflage to escape detection, using an explosive burst of flight on short wings as a last resort. Male pheasants (such peafowl) often have long trains of feathers and are brightly coloured.

PAINTED FRANCOLIN *Francolinus pictus*

Size: 31cm

Habitat: Grassland interspersed with scrub forest.

Distribution: In Sri Lanka confined in the East to area around Nilgala in Moneragala District.

Voice: Crackling sequence of *kruk! kuk kurra kuk kurra*. First note is short and abrupt, followed by two double-noted calls. Distinct call unlikely to be confused with that of Grey Francolin.

Status: Scarce Resident.

Rufous head. Brownish wings heavily patterned. Sexes similar but female has striking blackish-brown barring on white underparts. In male, underpart pattern is bolder and is more like large white spots forming bars on a black background. Larger size and heavy barring on underparts make confusion unlikely with plainer Grey Francolin (see p. 91).

GREY FRANCOLIN

Francolinus pondicerianus pondicerianus

Size: 33cm

Habitat: Dry scrub.

Distribution: Dry lowlands in area north of Puttalam, extending to Jaffna Peninsula. It is curious that such a gregarious bird has not spread to suitable habitats in the South.

Voice: Metallic, loud, double-noted call that is repeated.

Status: Uncommon Resident.

Pale sandy-brown overall, with wings having chequered pattern in muted browns. Fowl-sized; confusion unlikely with much smaller quails. Oval throat-patch is yellow with distinct edges.

RAIN QUAIL *Coturnix coromandelica*

Size: 18cm

Habitat: Grassland and farmland.

Distribution: Recorded in dry lowlands and highlands in suitable habitats.

Voice: Repeated *week-week*.

Status: Highly Scarce Migrant.

Male has boldly patterned face; three black stripes on each side of face and black throat-stripe join a thin black collar. Male also heavily marked with black streaks on underparts, with streaking fusing on breast to form a smudge. Both sexes have long white streaks edged with black on upperparts. Female has a plainer face than male, with black lines on face replaced by dull brown. Pale underparts are lightly streaked with brown. Sexes easily told apart by stronger facial patterns on male.

BLUE-BREASTED QUAIL
Coturnix chinensis

Size: 14cm

Habitat: Wet grassland and scrub.

Distribution: Present throughout Sri Lanka, but shy and scarce. Despite its wide distribution, very few birders have seen it. Look for it in grassy margins of tanks in places like Uda Walawe National Park.

Voice: Metallic, staccato *chrink-chrink* with rising intonation.

Status: Uncommon Resident.

Male (India)

Male

Male

JUNGLE BUSH-QUAIL
Perdicula asiatica

Size: 17cm

Habitat: Grassland and scrub jungles.

Distribution: In Sri Lanka confined to east in area near Nilgala.

Voice: Throbbing, slightly metallic, long drawn-out call, rising in pitch towards the end.

Status: Scarce Resident.

Male is unmistakable, with chestnut underparts, blue on crown, and sides of body and upperparts largely blue with some brown flight feathers. Face has white gorget bordered with black; black throat and white face divided by thin black line. Female is plainer, with off-white chin and throat, dark brown line from base of bill running below eye to behind ear-coverts and warm brown supercilium. Breast and flanks are buffish with irregular barring, fading into off-white in centre of belly.

Male is brick-red on underparts, chin and throat. Both sexes share boldly marked facial pattern with two brick-red lines, brown stripe through eyes and dark crown edged with white. Upperparts in both sexes marked with long white streaks. Female neatly barred with black against an off-white background on underparts.

CEYLON SPURFOWL
Galloperdix bicalcarata ℮

Size: 34cm

Habitat: It seems to need densely shaded forests or riverine forests.

Distribution: Found wherever decent-sized patches of wet zone forest occur, from lowlands to highlands. Curiously, I have never heard it at Horton Plains National Park though I have heard it at Hakgala. In dry zone, found along riverine forests.

Voice: Fortunately, it is very vocal as it is often heard but hardly ever seen. A series of ascending notes that begin as a series of yelping notes and change into a series of high-pitched *chuk chuk* notes. Sequence is broken by interspersal of differently pitched notes. Complex, varied but unmistakable.

Status: Uncommon Endemic.

Male

Female

Male boldly spangled in white edged with black. Large, teardrop markings on wings. Red facial patch, duller in female. Female dull brown overall. Always seen in pairs, and does not form flocks.

Male

Female

CEYLON JUNGLEFOWL
Gallus lafayetii ℮

Size: m-69cm/f-36cm

Habitat: Appears to survive only where sizeable tracts of protected areas remain. This could be an effect of hunting, and the birds are very shy outside national parks.

Distribution: Widespread up to the mountains wherever large tracts of forest survive. Dry zone seems to be its stronghold.

Voice: Call is a tremulous *churro-choik churro-choik* with an incredulous-sounding intonation.

Status: Common Endemic.

Female brown with barred wings. Yellow patch in middle of comb distinguishes male from domestic cockerel. Neck and mantle are golden.

INDIAN PEAFOWL *Pavo cristatus*

Size: m-110cm (2–2.5m w train); f-86cm

Habitat: Dry zone scrub jungles.

Distribution: Mainly in dry lowlands, though it is found in a few wet zone areas that border dry lowlands. There are many reports of it being seen from the Southern Expressway, which suggests that its range is expanding.

Voice: Honk followed by loud, double-noted, mournful-sounding, trumpeting call that rises slightly in pitch.

Status: Resident.

Male has iridescent, glossy 'peacock-blue' on head and neck in breeding plumage. Train of feathers is grown during the breeding season, when males dance with tail feathers raised to form a fan with shimmering eyes. Outside the breeding season the male sheds the feather train. Male retains its black-and-white barred wings and scapular feathers, whereas female's are brown with pale edges.

GRUIFORMES
This order includes families such as the familiar cranes and bustards. Only one family is found in Sri Lanka – the buttonquails, which some authors believe should be placed within the waders (order Charadriiformes).

Buttonquails Family *Turnicidae*
Buttonquails are small, cryptically coloured, round, quail-like birds. The quails are in a different order of birds.

SMALL BUTTONQUAIL
Turnix sylvaticus

Size: 13cm

Habitat: Grassland, scrubland and wooded thickets bordered by open country.

Distribution: Recorded in dry lowlands.

Voice: Long-drawn *boom*. Also a hollow sounding, throbbing call.

Status: Vagrant.

Broad white supercilium, white lores and brown streak extending from eye to behind ear-coverts. Pale irides. Spotted on breast with a diffuse buffish breast-band. Scapulars are broadly fringed in buff, creating lines. Mantle and coverts are also fringed in buff. Upperparts have scaled appearance. Smaller than other quails. In flight, buff wing-coverts on upper wing contrast with dark flight feathers.

BARRED BUTTONQUAIL (BARRED BUSTARD-QUAIL) *Turnix suscitator*

Size: 15cm

Habitat: Scrub jungle.

Distribution: Mainly seen in dry-zone scrub jungles, and occasionally in wet zone.

Voice: Similar to throb of a motorbike or low-frequency engine.

Status: Resident.

Male

Female

The role of the sexes is reversed in this species, and this is also reflected in the plumage. Female has a black chin and throat, and black-and-white barring on upper breast and sides of neck. Brighter than male. Males incubate the eggs and look after the young. Female lays a clutch of eggs and may move on to find another male to mate with, then lay another clutch of eggs.

Rails, crakes, gallinules and coots Family *Rallidae*
Rails and crakes have flattened body shapes – perhaps an adaptation to threading their way through dense thickets of reeds. Gallinules and coots have rounded bodies and frontal shields on their heads, and are prone to aggression.

SLATY-LEGGED CRAKE
Rallina eurizonoides amauroptera

Size: 25cm

Habitat: Marshes and damp thickets.

Distribution: Can be seen on arrival anywhere in Sri Lanka. Most records appear to be from the wet zone and within that in highlands, but this may be observer bias as the birding circuit focuses on the wet zone.

Voice: A repeated *onk onk*, with a nasal twang.

Status: Uncommon Migrant and Highly Scarce Resident. There is a record of a bird with chicks.

Slaty legs, ruddy breast and grey bill. Slaty-breasted Rail (see p. 96) also has slaty legs, but its breast is slaty and it has barred upperparts. Black-and-white barring on underparts of Slaty-legged is sharper and more contrasting. Upperparts are dark brown with no barring. Juveniles are similar, but juvenile Slaty-breasted has darker breast and neck and unbarred upperparts. Juveniles of both species have pale chins.

SLATY-BREASTED RAIL
Rallus striatus albiventer

Size: 27cm

Habitat: Marshy areas and paddy fields. Skulks in cover.

Distribution: Resident throughout Sri Lanka, but often overlooked due to its discreet habits. Resident population supplemented in winter by migrants.

Voice: Frog-like *blip* note, frequently repeated.

Status: Uncommon Resident and Uncommon Migrant.

Slaty legs (see also Slaty-legged Crake, p. 95). Slaty breast can look bluish. Pale reddish bill separates it from Slaty-legged, but confusion is possible with vagrant Brown-cheeked Water Rail (see right), which has a red bill, but does not have barred upperparts. Barring on underparts is also bolder. Juvenile is barred above.

BROWN-CHEEKED WATER RAIL
Rallus indicus

Size: 28cm

Habitat: Marshes. May be found in marshes adjoining paddy cultivation, but seems to be absent from those close to large cities.

Distribution: Most records are from 19th-century hunters in wet lowlands. Could potentially turn up in wetlands anywhere in Sri Lanka.

Voice: Metallic *trik-trik*, repeated a few times. Slightly tremulous, reminiscent of an amphibian's call.

Status: Vagrant.

Long red bill and red legs, brown upperparts, and slaty-blue on face and underparts up to belly, which is boldly marked with thick black-and-white bars. Confusion is possible with Slaty-breasted Rail (see left), but in Slaty-breasted legs are grey not red, barring is neater with thinner bars, and upperparts are lighter grey-brown and marked with thin white barring.

CORN CRAKE Crex crex

Size: 25cm

Habitat: Unlike other rails, prefers damp meadows close to water and is found in aquatic habitats. In Europe breeds in dry hayfields.

Distribution: Anywhere with damp meadows; most likely to turn up in lowlands.

Voice: Typically silent, but during breeding season utters a hoarse, mechanical, rasping call likened to *crex crex*. A cricket-like, repeated *creek creek*.

Status: Vagrant.

Bluish on face and neck with a brown mask; belly with heavy brown-and-white barring, with some of brown bars edged in black. Wings are brown. Upperparts paler brown with dark centres to feathers arranged in thick lines that run along body. Long legs trail behind tail in flight.

WHITE-BREASTED WATERHEN
Amaurornis phoenicurus phoenicurus

Size: 32cm

Habitat: Wetlands, canals, ditches and similar places, in lowlands and mid-hills.

Distribution: Widespread throughout Sri Lanka. No patch of marsh even in suburban areas appears to be without a pair of these birds.

Voice: A pair of birds may engage in some noisy duetting with the call *korawak korawak* repeated at length. Sinhala name for the species, *Korawakka*, is onomatopoeic.

Status: Common Resident.

Prominent white face and breast, and slaty upperparts. Young are covered in black down.

Eastern Baillon's Crake
Porzana pusilla

Size: 19cm

Habitat: Prefers marshland and other aquatic habitats that are reed fringed.

Distribution: May turn up anywhere in Sri Lanka and take up residence in a wetland. One disoriented bird arrived in my school in Colombo.

Voice: Drawn-out, rapid rattle that changes in pitch.

Status: Highly Scarce Migrant.

Ruddy-breasted Crake *Porzana fusca*

Size: 22cm

Habitat: Wide range of wetland habitats from marshes and paddy fields, to floodplains of rivers.

Distribution: Found throughout Sri Lanka, but probably occurs in highest numbers in wet lowlands.

Voice: Drawn-out, rapid rattle, reminiscent of call of Little Grebe (see p. 34).

Status: Uncommon Resident.

Small crake that looks like a miniature Water Rail with brown upperparts, bluish face and underparts, and bold black-and-white barring on belly. Unlike the much larger Brown-cheeked Water Rail (see p. 96), it has a short bill and slaty legs. Wings and back marked with black splotches with white edging.

Chestnut on face and underparts up to belly. White barring on chestnut belly. Barring continues on undertail-coverts. Red legs and grey bill. Red legs and absence of black-and-white barring on belly easily separate it from Slaty-legged Crake (see p. 95).

WATERCOCK *Gallicrex cinerea cinerea*

Size: m-43cm/f-36cm

Habitat: Secretive bird of lowland marshes. Most likely to emerge into the open in the evenings, though I have at times seen it in the heat of the afternoon walking across a lily-covered lake. Talangama Wetland is a regular site for birders.

Distribution: Widespread throughout lowlands. Prefers marshes bordered by dense vegetation.

Voice: Repeated, soft and liquid-like, sharp *ooh*.

Status: Uncommon Resident.

Breeding male is dark (almost black) overall, with red 'plate' on upper mandible. Non-breeding male and female are a similar pale brown overall. Upperparts are scaly, with pale edges to dark-centred feathers.

Male

PURPLE SWAMPHEN
Porphyrio porphyrio poliocephalus

Size: 43cm

Habitat: Common in freshwater marshland in lowlands.

Distribution: Found throughout lowlands. Most common in dry zone as a result of the larger number of open waterbodies there. Feeds largely on aquatic plant material.

Voice: Grating calls, metallic trilling calls. Also nasal *wah wah* calls like a tired animal exhaling breath.

Status: Resident.

Large rail with red legs and bill, and blue body. Juveniles are grey and downy. Sexes are similar. Males are highly combative during the breeding season and frequent clashes take place between them. Occasionally subject to illegal hunting by local people as a substitute for poultry. Can be confiding where not subject to human threats.

Common Moorhen
Gallinula chloropus indica

Size: 42cm

Habitat: Freshwater bodies in lowlands.

Distribution: Throughout lowlands. Not particularly common anywhere.

Voice: Utters a few croaking notes.

Status: Resident.

Overall dark-looking bird, on closer inspection with brownish wings and dark underparts. White jagged line on flanks shows as horizontal line when bird is floating in water in duck-like fashion. Yellow-tipped bill with red base and forehead plate. Dark underparts distinguish it from White-breasted Waterhen (see p. 97). Shorter-legged than Watercock (see p. 99), which lacks white line.

Eurasian Coot *Fulica atra atra*

Size: 42cm

Habitat: Lakes and ponds. Prefers open water, though it also occurs on lily-covered ponds.

Distribution: Dry zone, mainly in north of Sri Lanka. Migrants turn up in west and south-east.

Voice: Metallic *kreek kreek* call. Another call is a sharp, metallic, single-note *chink*, like two pebbles being struck together.

Status: Uncommon Resident and Scarce Migrant.

Black bird with white bill and white plate on forehead. Juveniles are greyish and covered in downy feathers. Mainly a migrant, and most birds seen are in adult plumage. Has been recorded breeding in Mannar and Annaiwilundawa Wetland in north-west of Sri Lanka.

CHARADRIIFORMES
This diverse order contains 18 families, many of which have little physical resemblance to each other.

Jacanas Family *Jacanidae*
Long toes distribute the body weight of these birds, enabling them to walk on floating vegetation.

PHEASANT-TAILED JACANA
Hydrophasianus chirurgus

Size: 31cm w/o streamers

Habitat: Frequents lowland marshes and lily-covered lakes.

Distribution: Most common in dry zone.

Voice: Liquid, bubbly *toonk toonk* calls. Also various mewing and screechy, grating calls.

Status: Resident.

Non-breeding

Breeding

Non-breeding adults similar to juveniles, but show trace of yellow on neck. Breeding birds have showy, long tails. Sexes look similar, but female is slightly bigger than male, with longer toes – a reversal of the usual size dimorphism of the sexes.

Painted-snipes Family *Rostratulidae*
These are snipe-like birds, more strikingly patterned than snipes, with the role of the sexes reversed.

Male

Female

GREATER PAINTED-SNIPE
Rostratula benghalensis benghalensis

Size: 25cm

Habitat: Found in marshes and paddy fields in lowlands. Often overlooked due to its discreet habits. Generally nocturnal, but sometimes seen feeding during the day.

Distribution: Found throughout lowlands.

Voice: Repeated, single-note *wonk*. Very amphibian-like in quality and may be mistaken by many people for a frog calling.

Status: Uncommon Resident.

Female brighter coloured than male – an example of reversal of the usual sexual dimorphism. Body has a prominent white band that curves over wings in an arc. Eyes encircled by a white ring with a line extending to back, like a white horizontal 'comma' sign around eye. Bill shorter than a snipe's and slightly downcurved, unlike bills of snipes. Male has yellowish barring on wing and is more patterned than female.

Oystercatchers Family *Haematopodidae*
These waders have a plumage that is black or sooty overall, made up of a combination of black and white. All oystercatchers have long red bills, and long red or pink legs.

Plovers Family *Charadriidae*
Plovers are round-headed, dumpy-looking, ground-dwelling birds with short, stubby bills.

Many of them, like Kentish Plovers, give the appearance of lacking a neck, with the head looking as though it has been fused directly to the body.

EURASIAN OYSTERCATCHER
Haematopus ostralegus longipes

Size: 42cm

Habitat: Lowland marshes and wet habitats.

Distribution: Can turn up anywhere on coasts, on beaches and estuaries. Known localities include Bundala, Chilaw Sand Spit and Mannar Causeway.

Voice: High-pitched, loud *tleep* note repeated quickly when excited, and more slowly when making contact.

Status: Scarce Migrant.

Distinctive, large, black-and-white wader with long red bill and long red legs. Black head and breast and upperparts, with white underparts. In flight shows a broad white wing-bar. Juvenile duller than adults on upperparts.

PACIFIC GOLDEN PLOVER Pluvialis fulva

Winter   Summer

Size: 24cm

Habitat: Wet pastures and short-cropped grasslands in lowlands. Habitat preference is different from that of Grey Plover.

Distribution: Found throughout lowlands on coast or close to it. Will occupy wet meadows, but usually those within a few kilometres of the coast.

Voice: Repeated, high, fluty notes, *tee too whee-oh*.

Status: Migrant. Occurs in small flocks.

Buff to golden tips on mantle feathers. These tips occupy about half of the visible length and are heart shaped on close examination. In non-breeding plumage, Pacific Golden Plovers and Grey Plovers (see p. 103) are superficially similar, but Grey Plover is greyer and always has the diagnostic black axillaries. In breeding plumage, bases of wing feathers turn black and golden tips contrast strongly, making evident the name 'golden plover'.

Advanced ID notes: Pacific Golden Plover is distinguished from European Golden Plover *P. apricaria* by white instead of grey-brown axillaries and inner-wing lining. Toes project beyond tail in Pacific but not in European. American Golden Plover *P. dominica* is not likely to be seen in Sri Lanka, but it too has brown axillaries and wing lining. It is a vagrant to Europe, where this feature is useful in distinguishing it from European Golden.

GREY PLOVER *Pluvialis squatarola*

Size: 31cm

Habitat: In small numbers in mudflats on estuaries, lagoons and salt pans.

Distribution: Through coastal strip, mainly in dry lowlands.

Voice: Double-noted, high-pitched *pee wee* call.

Status: Uncommon Migrant.

Summer

Black axillaries (armpits) are diagnostic in all plumages. In winter, dark mantle feathers have pale, almost white edges. Lacks buff to golden tips on mantle feathers of Pacific Golden Plover (see p. 102).

Winter

Winter

Summer

Winter

COMMON RINGED PLOVER
Charadrius hiaticula

Size: 19cm

Habitat: Coastal mudflats in dry lowlands.

Distribution: May turn up anywhere on coastal wetlands and mudflats. Most likely to be seen in area north of Mannar to Northern Peninsula.

Voice: Repeated, high, musical *phlew* note. Repeated more frequently when birds are interacting.

Status: Scarce Migrant.

Distinct white wing-bar separates Common from Little Ringed Plover (see p. 104). It is also larger, with orange base to black bill. Breeding adult lacks yellow eye-ring of breeding Little Ringed. Juveniles are similar, but juvenile Little Ringed may show traces of a yellow orbital ring. In Little Ringed, brown band below eye has a sharp downwards point; in Common Ringed, bottom border is smooth and curved. However, this distinction is not apparent in the very early stages. In juvenile Little Ringed, tertials extend over primary tips. Bill in Common Ringed is stouter.

LITTLE RINGED PLOVER
Charadrius dubius curonicus

Size: 17cm

Habitat: Open areas, especially mudflats in dry lowlands. At times, on ploughed paddy fields.

Distribution: Migrant birds appear in wet zone.

Voice: Repeated, high-pitched, tremulous *phew* note.

Status: *C. d. curonicus* is an Uncommon Migrant and *C. d. jerdoni* is an Uncommon Resident.

Yellow eye-ring and very thin white wing-bar distinguish this species from Common Ringed Plover (see p. 103). Note that in juvenile, eye-ring is dull. See also notes under Common Ringed Plover for distinguishing juveniles. In breeding plumage, eye-ring is bright and swollen. Prominent black breast-band across white underparts, oval black patch below eye and black band on crown behind white forehead.

Breeding

Breeding

Breeding

Non-breeding

KENTISH PLOVER *Charadrius alexandrinus*

Size: 17cm

Habitat: Mud flats, sand banks and short-cropped grassland in coastal areas.

Distribution: All around coastal regions, but absent where coastline is developed.

Voice: Whistled note comprising a short *chwitz*; sounds more like a passerine.

Status: Race *C. a. alexandrinus* is an Uncommon Resident and race *C. a. seebohmi* is an Uncommon Migrant.

Smallest of plovers seen in Sri Lanka. White underparts and sandy-brown upperparts. In breeding plumage, develops small black wedge adjoining corner of wing. In non-breeding plumage, this becomes brown and indistinct. Can be distinguished from Lesser Sand Plover (see p. 105) by smaller size and slimmer bill, which is more pointed. Head is flatter on crown. Develops rust-brown on crown in breeding plumage.

LESSER SAND PLOVER
Charadrius mongolus atrifrons

Size: 19cm

Habitat: Open, short-cropped grassland in dry lowlands. Most abundant near coast, where it also frequents tidal habitats. In wet zone it can turn up on wet meadows. A flock regularly occupies the Kotte Marshes near Colombo.

Distribution: Throughout lowlands, but mainly in dry lowlands.

Voice: High-pitched, piping notes interspersed with tremulous notes.

Status: Migrant.

Winter

Winter

Winter

During most of its stay in Sri Lanka this is a nondescript, sandy-brown plover with white underparts. Often congregates in large flocks. Shorter and small billed compared with Greater Sand Plover (see right). Develops black mask and orange-buff breast-band in breeding plumage.

GREATER SAND PLOVER
Charadrius leschenaultii leschenaultii

Size: 22cm

Habitat: Estuarine mudflats in dry lowlands. Also seen on beaches, but for feeding moves into areas swept by the tide.

Distribution: Disperses around coastline after arrival. Northern half Sri Lanka may be the best place to see it, as it has large areas of coastal estuaries.

Voice: Rapid, tremulous, high-pitched notes.

Status: Scarce Migrant. Compared with Lesser Sand Plover, appears in relatively small numbers.

Black mask and orange-buff breast-band in breeding plumage. Heavier-billed than Lesser Sand Plover (see left) and has a longer-legged appearance. Legs are pale green. To some, Greater Sand Plover may appear a bit top heavy as it is longer legged than Lesser and appears to be supporting a greater body mass. It takes time and practice in looking at the more common Lesser Sand Plover to gain a feel for telling the two species apart. Greater prefers coastal habitats, which are estuarine, and is usually in ones and twos. This also provides a clue to its identity.

CASPIAN PLOVER *Charadrius asiaticus*

Size: 19cm

Habitat: Mudflats and lagoon edges in dry lowlands.

Distribution: May turn up anywhere in coastal habitats.

Status: Scarce Migrant. A few records at the most each year.

Winter

Summer

Winter

Summer

In non-breeding plumage, similar to Oriental Plover (see right). Legs are grey-green and white wing-bar is more distinct than faint wing-bar in Oriental Plover. Underwing is pale. May be seen with flocks of Lesser Sand Plover, from which it can be distinguished by its longer legs, sharp, pointed bill and thick white supercilium extending behind eye. Legs do not project beyond tail in flight. In breeding plumage, orange breast-band with dark line on lower border; white face has a dark patch between eye and ear-coverts.

ORIENTAL PLOVER *Charadrius veredus*

Size: 24cm

Habitat: Mudflats and coastal plains.

Distribution: Most likely to occur in coastal areas in dry lowlands.

Voice: Repeated piping *chip*. Most rapid when uttered in flight.

Status: Vagrant.

In winter plumage, can be distinguished from similar Caspian Plover (see left) by lack of wing-bar, or having only a faint one. Long yellow legs project beyond tail in flight, and underwing is dark (pale in Caspian). In breeding plumage, birds have white head and neck, and orange breast-band with a thick black border on belly.

YELLOW-WATTLED LAPWING
Vanellus malabaricus

Size: 27cm

Habitat: Open areas in dry lowlands. Prefers dry, short-cropped grassland. Does not occur on damp meadows as does Red-wattled. Feeds on animals such as invertebrates, worms and insects, and hunts by picking or pulling prey from the ground and surface vegetation.

Distribution: Dry lowlands, especially in arid-zone areas.

Voice: Main call is a repeated, drawn-out, screechy single note.

Status: Uncommon Resident.

Sexes are similar, with both having pronounced yellow lappets on face. Dark cap bordered with white below. Breast brown and not black as in Red-wattled Lapwing (see p. 108).

Juvenile

GREY-HEADED LAPWING
Vanellus cinereus

Size: 37cm

Habitat: Wet meadows and other habitats with short grassland.

Distribution: May potentially arrive anywhere in lowlands, though most records are from dry lowlands.

Voice: High-pitched *kee kee*, at times a little tremulous. Reminiscent of call of Black-winged Stilt (see p. 127).

Status: Vagrant.

Grey head and neck bordered by broad black breast-band and greyish-brown wings make it unlikely that this species will be confused with any other plover. Upperwing has black outerwing, and inner forewing is grey-brown with a broad white triangle on the secondaries. Black tail contrasts with white rump uppertail-coverts.

RED-WATTLED LAPWING
Vanellus indicus lankae

Size: 33cm

Habitat: Wet habitats. Will occupy small wetland patches even in cities.

Distribution: Common throughout Sri Lanka.

Voice: Call likened to 'did he do it', which is an onomatopoeic common name.

Status: Common Resident.

Winter

SOCIABLE PLOVER
Vanellus gregarius

Size: 33cm

Habitat: Damp grassland and other short, grassy habitats, including playing fields.

Distribution: Most early records were from wet lowlands, with a record of a flock in Wilpattu National Park.

Voice: Harsh *chark-chark-chark* flight call.

Status: Vagrant.

Brown upperparts with white underparts and sides of face and neck. White contrasts strongly with black cap, nape and breast. Legs yellow as in Yellow-wattled Lapwing (see p. 107). Overall impression is of a black-and-white plover with red on lores and base of bill.

Conspicuous pale supercilium in all plumages. In adult breeding plumage, crown turns black with a thick black line from bill to behind ear-coverts. In winter, loreal area is pale and crown is brown. In breeding plumage, belly turns black with rear edge rufous. Belly turns pale in winter plumage.

> **Sandpipers and allies** Family *Scolopacidae*
> A wide variety of waders comprise this family. Leg height and bill structure are very varied, allowing different species to occupy different niches in aquatic habitats. Most species in their wintering quarters use estuaries and mudflats.

EURASIAN WOODCOCK
Scolopax rusticola

Size: 36cm

Habitat: Damp meadows and wetland habitats. Hard to see unless it is flushed.

Distribution: Arrives throughout Sri Lanka.

Voice: Sharp *tseep*.

Status: Highly Scarce Migrant.

Head has a unique pattern with transverse dark brown-and-white bands on hindcrown, dirty white forecrown, dirty white face with a dark loreal line continuing slightly through eye, and dark line on cheek. Longer bill has a broad base and seems chunkier than bills of snipe species. Upperparts richly patterned in browns, with scapulars having arrowhead markings. Pot-bellied.

WOOD SNIPE *Gallinago nemoricola*

Size: 30cm

Habitat: Damp meadows. May turn up in short, periodically inundated grassland adjoining lakes.

Distribution: Most records have been from 19th-century hunters in highlands, but could turn up anywhere in Sri Lanka.

Voice: Harsh, punchy *cheep-cheep chouk chouk* that is nasal. Reminiscent of a reed warbler.

Status: Vagrant.

Large, dark and stocky compared with Pintail and Great Snipes (see pp. 110 and 111). Lacks areas of white on underparts, being completely barred on belly. Dark underwing. Very little white on corners of tail, unlike in Common Snipe (see p. 111) and Great Snipe. No white trailing edge to wing, which is present in Common and Great.

PINTAIL SNIPE *Gallinago stenura*

Size: 26cm

Habitat: Paddy fields and marshes in lowlands.

Distribution: Spreads throughout Sri Lanka in marshy habitats, all the way to highlands.

Voice: When flushed, takes off with a harsh *kreik* call. Not guttural as in herons.

Status: Migrant.

The most common of the snipes – almost all of the snipe seen in Sri Lanka are Pintail Snipe. Common Snipe (see p. 111) is a scarce migrant to Sri Lanka and is best distinguished from this species by white trailing edge to secondaries in flight. Even in the field, Common is noticeably longer billed to someone familiar with Pintail Snipe. In Pintail, buff stripe over eye is wider than black eye-line, where the two join base of bill, but this is not easy to see in the field. Underwing in Common is paler at wing-base.

SWINHOE'S SNIPE
Gallinago megala

Size: 28cm

Habitat: Wet meadows, paddy fields and marshes.

Distribution: The few records are from wet lowlands, but it may occur anywhere in wetlands.

Voice: Contact calls of Pintail and Swinhoe's Snipes are similar – a sharp *kreek*. Song is distinctive. Swinhoe's song is a wheezy, piping rattle that rises to a crescendo and abruptly falls away.

Status: Vagrant.

Very similar to Pintail Snipe (see left) but with longer bill. Crown-stripe does not usually reach bill. Toes barely project beyond tail. Legs generally thinner than those of Pintail. However, separation of Pintail from Swinhoe's is difficult, and the only reliable feature, if seen, is the pin-shaped tail feathers in Pintail. The song, if heard (unlikely in Sri Lanka), is the best clue.

GREAT SNIPE *Gallinago media*

Size: 28cm

Habitat: Wetland habitats from paddy fields and marshes to damp meadows.

Distribution: The few records have been from wet lowlands in the West, but could occur anywhere in Sri Lanka on wetlands.

Voice: Song is a thin, tinkling call accompanied by a rattle as if a marble has been dropped on a floor.

Status: Vagrant.

Distinguished from Common Snipe (see left) by extensive barring on belly, with no unbarred white belly area as in Common. White wing-bars more prominent than in any of the other snipe species recorded in Sri Lanka. White trailing edge less distinct than in Common. White on corner of tail conspicuous and more defined than in the other snipe species. Dark underwing. In Common, underwing has pale bands.

COMMON SNIPE
Gallinago gallinago gallinago

Size: 26cm

Habitat: Wetlands. I once photographed a pair in the Kotte Marshes, which is a disturbed site close to Columbo. It is worth examining all snipe carefully.

Distribution: Recorded in lowlands. Absence of records in highlands is puzzling.

Voice: When flushed, takes off with a harsh *kreik* call. Similar to Pintail Snipe's call.

Status: Uncommon Migrant. May be overlooked for more common Pintail Snipe.

See Pintail Snipe (see p. 110) for more notes on identification. Has a white trailing edge to wings and broad white bars on underwing. White edge to scapulars is broader on outer edges.

JACK SNIPE Lymnocryptes minimus

Size: 21cm

Habitat: Wetland habitats including paddy fields and marshes, and also where waterbodies are fringed with vegetation.

Distribution: Occurs throughout Sri Lanka.

Voice: Song is a throaty, gargling sound.

Status: Highly Scarce Migrant.

Short-legged, flat-profiled snipe with habit of bobbing. Shorter-billed than snipe. Split supercilium. Dark line through eye and another dark, curved line on cheeks. Sometimes lines join to form a crescent. Conspicuous yellowish, parallel lines in scapulars and mantle. Streaked on breast; white belly.

Winter

Winter

BLACK-TAILED GODWIT Limosa limosa

Size: >46cm

Habitat: Mudflats, estuaries and lagoons in lowlands. Occasionally in marshes and paddy fields on passage. Most common in dry lowlands.

Distribution: Found throughout lowlands in coastal habitats. May be encountered in wetlands that are inland, but usually no more than a few kilometres from the sea.

Voice: Fast, repeated *wee-oh* that is thin and sounds urgent.

Status: Migrant. Race *L. l. melanuroides*, known as Eastern Black-tailed Godwit, is a Highly Scarce Migrant.

White wing-bar and black tail-tip are diagnostic and distinguish this species easily from similar Bar-tailed Godwit (see p. 113). Occurs in large flocks. In non-breeding plumage, superficially similar to Bar-tailed, but longer legs, more uniformly grey upperparts and straight bill help to distinguish it. In breeding plumage, body turns rufous. Mantle and flight feathers have black centres notched with rufous.

BAR-TAILED GODWIT
Limosa lapponica lapponica

Size: 39cm

Habitat: Similar to Black-tailed. Usually seen singly.

Distribution: Coastal wetlands. Most records are from area from Mannar to Northern Peninsula.

Voice: High-pitched *keek keek* calls reminiscent of those of Black-winged Stilt (see p. 127). Song a rapidly repeated single note.

Status: Scarce Migrant.

Winter

Acquiring winter

Winter

Shorter legs and streakier in appearance than Black-tailed Godwit (see p. 112). Upturn in bill is subtle but noticeable in field. Mantle and flight feathers have dark shaft lines, giving it a streaky appearance. In breeding plumage more rufous on underparts than Black-tailed.

WHIMBREL *Numenius phaeopus*

Size: 43cm

Habitat: Coastal mudflats, mangroves and brackish water habitats. Sometimes feeds on beaches in areas washed by waves. Curlew is seldom seen on beaches.

Distribution: Can show up anywhere on the coastline, especially on beaches. Most common in dry lowlands, but this may be a result of habitat destruction on wet zone coastline.

Voice: Rapidly repeated, piping notes.

Status: Uncommon Migrant. Two races, *N. p. phaeopus* and *N. p. variegatus*, have been recorded.

Can be distinguished from similar Eurasian Curlew (see p. 114) by smaller size and shorter bill. Curlew's bill is noticeably longer, and Whimbrel has median stripe on crown.

EURASIAN CURLEW
Numenius arquata orientalis

Size: 58cm

Habitat: Mudflats and meadows beside coasts. Seems to favour freshwater meadows and damp grassland, where it can probe the ground for worms and other invertebrates. Does not like the brackish habitats that the Whimbrel favours. However, even when seen inland on damp grassland, it is always within a few kilometres from the sea.

Distribution: Coastal areas.

Voice: Liquid, bubbling calls rising in intonation. Last note is likened to 'curlew', hence the onomatopoeic name curlew.

Status: Uncommon Migrant.

The only possible confusion is with the smaller-sized and shorter-billed Whimbrel (see p. 113). An elegant wader that literally stands tall among other waders.

Winter

Winter

SPOTTED REDSHANK
Tringa erythropus

Size: 33cm

Habitat: Winters in estuaries, mudflats and low-lying coastal areas. Breeds on bogs, marshes and Arctic Taiga.

Distribution: Winters in coastal mudflats and salt pans in South, but may turn up anywhere in suitable coastal areas.

Voice: Loud *tsweet tsweet*, a clacking call followed by a musical piping note with a hint of quivering. Common Redshank's calls are purer and lack the quiver in Spotted Redshank's calls.

Status: Vagrant.

In winter, similar to Common Redshank (see p. 115), but upper mandible is black with base of lower mandible red. In breeding plumage, head, neck and underparts turn black with white spots on mantle and coverts. In all plumages, absence of a white wing-bar is a key diagnostic feature distinguishing it from Common Redshank.

COMMON REDSHANK *Tringa totanus*

Size: 28cm

Habitat: Estuaries, lagoons and other brackish habitats. On passage, in freshwater habitats like marshes and paddy fields. Most common in dry lowlands.

Distribution: Mainly in coastal areas, where it occupies a mix of wetland habitats, from freshwater marshes to mudflats and mangroves. Wintering birds tend to stop over only for brief periods in freshwater habitats in wet zone. In dry zone, they occupy flooded fields and marshes that are not far from the coast.

Voice: Liquid, melodious *peeuw* of medium length instantly identifies this species. For many birders this call is the audio signature of a wader habitat. The birds are very vocal, and in England were known locally as 'the wardens of the marshes', as they are quick to alert others of intruders.

Status: Common Migrant. Two subspecies, *T. t. eurhinus* and *T. t. terrignotae*, have been recorded.

Summer

Red legs and white wing-bar on upperwing make it easy to identify this wader. Similar looking to Spotted Redshank (see p. 114); a very scarce migrant. In flight, absence of white wing-bar in Spotted makes identification easy. At rest, Spotted and Common Redshanks are similar, but in Spotted red on base of bill is confined to lower mandible. However, juvenile Common Redshanks sometimes have reddish base confined to lower mandible, so wing-bar is a safe diagnostic feature. In breeding plumage, Spotted Redshank is very different, with a black head, neck and underparts, and the wings and mantle feathers turning black with pale edges.

Winter

COMMON GREENSHANK *Tringa nebularia*

Size: 28cm

Habitat: Paddy fields, estuaries, lagoons and similar areas in lowlands.

Distribution: Prefers brackish wetlands. Also uses freshwater wetlands. Stops over in paddy fields and marshes on migration to southern dry zone coastal wetlands.

Voice: Single *kip kip* notes uttered in rapid sequence. In flight a double-noted *chew-chew*. Calls a little melodious.

Status: Uncommon Migrant.

Bill heavier and slightly upturned compared with that of Marsh Sandpiper (see p. 116). Heavier build. Bill often the best field character. See also notes under Marsh Sandpiper. Flight call is diagnostic. Juvenile has browner upperparts than adults, with pale edges. Adult winter is grey with mantle feathers fringed with dark-and-pale double band. Tertials show traces of notched edges, which become conspicuous in breeding plumage. Head and neck become streaked in summer plumage.

MARSH SANDPIPER *Tringa stagnatilis*

Size: 25cm

Habitat: Marshes, paddy fields, estuaries, lagoons and similar areas in lowlands.

Distribution: Spreads around coastal habitats. Being one of the taller waders, it often wades deep into pools. Marsh Sandpipers sometimes spend a few days in freshwater marshes and paddy fields as they travel to and from their preferred coastal wetlands in dry lowlands.

Voice: High-pitched, strident *pip pip*.

Status: Common Migrant.

Summer

Winter

Pencil-thin bill, slender build and pale overall colour help identify this wader. Juvenile has dark mantle and flight feathers. In winter plumage, adults are plain with white head and neck. Wing feathers are pale edged. In breeding plumage, head and neck are flecked to form streaks. Wing and mantle feathers become light brown, and middle of mantle feathers has black markings (which in close-up look like the silhouette of a bird of prey in flight). In flight, Marsh Sandpiper shows a long, narrow white wedge on upperparts; this is wide in Common Greenshank (see p. 115). Also in Marsh, toes project well beyond tail. Leg colour varies from greenish to yellow.

Winter

Winter

GREEN SANDPIPER *Tringa ochropus*

Size: 24cm

Habitat: Prefers freshwater pools in lowlands.

Distribution: Found throughout lowlands, but seems to prefer dry lowlands. Occasionally seen in streams and other wet habitats in highlands.

Voice: Can make a series of rapid *pip pip* notes followed by longer fluty notes. Flight call is 2–3 *thchew thchew* notes, which are clean and clear.

Status: Uncommon Migrant.

White supercilium does not extend behind eye (see also Wood Sandpiper, p. 117). In flight, dark underwings give it a black-and-white appearance at a distance.

WOOD SANDPIPER *Tringa glareola*

Size: 21cm

Habitat: Marshes, paddy fields, estuaries, lagoons and similar areas in lowlands.

Distribution: Spreads all over lowlands in marshy habitats. A few birds make their way up to the hills as well, but this is not typical. Most Wood Sandpipers seen in wet zone seem to stop over on passage to wetlands in southern half of Sri Lanka.

Voice: Flight call reminiscent of flight call of Common Sandpiper, but less thin and more liquid-like in quality. Other calls include piping notes, repeated frequently.

Status: Migrant.

Pale underwings (Green Sandpiper, see p. 116, has dark underwings). Pale supercilium extending beyond eye also distinguishes it from Green. Heavily speckled upperparts distinguish it from Green Sandpiper and Common Sandpiper (see pp. 116 and 118). Greenish-yellow legs, taller than in Common.

Winter

Winter

Winter

Winter

TEREK SANDPIPER *Xenus cinereus*

Size: 24cm

Habitat: Coastal mudflats, salt pans and estuaries in dry lowlands.

Distribution: Confined to mudflats and mangroves on the coast. Could turn up anywhere where there is suitable habitat. Numbers are higher in coastal wetlands in northern half of Sri Lanka.

Voice: Melancholy whistled or piped notes of medium length. Sometimes preceded by rapid double note.

Status: Uncommon Migrant.

Short-legged wader with a long, upturned bill. Distinctive profile. Legs are yellowish, though leg colour may be masked by mud. Bill has diffused yellow patch at base. Greyish-brown upperparts, with faint edges to feathers. Looks uniformly coloured in the field. Juveniles have thick edges to wing feathers.

COMMON SANDPIPER *Actitis hypoleucos*

Size: 36cm

Habitat: Marshes, paddy fields, estuaries, lagoons and similar places, mainly in lowlands. Sometimes along streams and canals.

Distribution: Anywhere in Sri Lanka. In highlands usually seen near watercourses.

Voice: High-pitched *tsee tsee tsee* notes, usually uttered in flight.

Status: Migrant.

Winter

Winter

'White finger' curving around bend of wing is a useful field identification characteristic. In flight, stiff-winged and shows a clear white wing-bar. Habit of bobbing tail.

RUDDY TURNSTONE
Arenaria interpres interpres

Winter

Summer

Size: 22cm

Habitat: Coastal mudflats, estuaries and similar places.

Distribution: On arrival spreads around coastal areas. Never far from the sea.

Voice: Tremulous, bubbly call that is short in duration.

Status: Migrant.

Plump and long body, low profile and broad-based bill make this wader easy to pick out. Body shape alone is sufficient to tell turnstones apart from other waders. In non-breeding plumage, upperparts are dark with dark breast-band and dirty white head. In summer, mantle turns chestnut, and throat, breast and eye-patch turn jet-black. White band between mantle and neck breaks up black on neck. This seems to work as disruptive colouration since if the birds are motionless, they can be hard to pick out.

ASIAN DOWITCHER
Limnodromus semipalmatus

Size: 34cm

Habitat: Estuaries and coastal lagoons.

Distribution: May potentially occur anywhere where coastal intertidal habitats remain undisturbed.

Voice: Flight call is a tremulous *kraark*. Slightly nasal.

Status: Vagrant.

Winter

Summer

Relatively large, stocky wader. Stout, straight grey bill with swollen tip. Looks short legged compared with, say, Black-tailed Godwit (see p. 112). Diffuse wing-bar and grey tail. In winter plumage is grey and brown. In summer plumage underparts and head turn rufous.

Winter

Winter

GREAT KNOT *Calidris tenuirostris*

Size: 27cm

Habitat: Mudflats, estuaries and lagoons.

Distribution: Recorded around coast of Sri Lanka, mainly where large concentrations of shorebirds occur in good-quality intertidal habitats.

Voice: Typically silent, with an occasional soft *prrt*.

Status: Highly Scarce Migrant.

Wings extend beyond tail giving attenuated look. In Red Knot (see p. 120), wings extend up to tail. Compared with Red Knot, longer legged, with bill longer and slightly downcurved, though in some birds bill looks straight and dagger-like. In winter more prominently marked with heavy spotting on breast and flanks. In breeding plumage, scapulars turn reddish, and black markings on breast and flanks are more pronounced.

RED KNOT Calidris canutus

Size: 24cm

Habitat: Mudflats, estuaries and other intertidal habitats.

Distribution: Anywhere on coast where large intertidal habitats remain.

Voice: Flight call is a repeated, high-pitched *whee-whee*. Song is a long-drawn series of notes, *too-wee too-wee*, with a change of pace; melancholy and musical.

Status: Vagrant.

Shorter, straight bill distinguishes this species from Great Knot and Ruff (see pp. 119 and 126). Latter has longer, noticeably downcurved bill. Shorter legged than Great Knot. Overall impression is of a plump, short-legged wader. Closed wings extend to tail; in Great, wings extend beyond tail, giving attenuated look. In winter, lacks coarse streaking on breast seen in Great. Note Ruff has scaly look with pale edges to mantle feathers. In summer, underparts turn brick-red and mantle feathers are red with black tips and central shafts.

Winter

Acquiring summer

Winter

Winter

SANDERLING Calidris alba

Size: 19cm

Habitat: Bird of the coastline, where undisturbed and relatively unpolluted stretches can be found.

Distribution: Can turn up anywhere on sandy beaches. Chilaw Sand Spit is a reliable site.

Voice: Sharp, slight tremulous *zit* call.

Status: Uncommon Migrant.

Short black legs and short black bill. In winter looks all white with black patch on bend of wing. In winter plumage, black shoulder-patch is concealed at times. White wing-bar contrasts strongly with black trailing edge of wing and dark greater coverts forming a dark edge to front of white wing-bar. Juveniles have darker mantle than adults. In breeding plumage, head, neck, mantle, scapulars and tertials acquire pale rufous tinge. Mantle feathers have thick black tips. Habit of running up and down as waves break on shore. Usually seen in flocks of around 7–15 birds.

LITTLE STINT *Calidris minuta*

Size: 13cm

Habitat: Estuaries and mudflats in lowlands. Most common in dry lowlands.

Distribution: Coastal areas, especially in dry zone.

Voice: High-pitched, rapidly repeated and rising *tseep tseep tseep.*

Status: Common Migrant.

Winter

Winter   Summer

One of the smallest waders. Bill black and blunt tipped. White underparts. In winter plumage (which birds are in during most of their time in Sri Lanka), upperparts are greyish, with mantle feathers having black shafts. In breeding plumage black shaft-streaks widen into black centres to feathers with thick rusty edges. Tertials become black-centred with rusty-red edges.

Acquiring summer

Summer

RUFOUS-NECKED STINT
Calidris ruficollis

Size: 14.5cm

Habitat: Mainly intertidal habitats in estuaries and lagoons. May be found in brackish water, ponds and salt pans near coast.

Distribution: Coastline; most records are from the South.

Voice: Repeated, shrill *tree.*

Status: Vagrant.

In breeding plumage, red on neck helps distinguish this species from Little Stint (see left), though breeding Little and Rufous-necked Stints both acquire red on face. In breeding Rufous-necked, mantle turns rufous with dark centres. Breeding Little is similar. Red neck with throat-streaks below it is thus the best identification feature. Juvenile Rufous-necked has black-centred mantle feathers restricted to top, contrasting with pale upperparts. Scapulars and wing-coverts extensively black-centred in juvenile Little.

TEMMINCK'S STINT Calidris temminckii

Size: 14cm

Habitat: Intertidal habitats as well as sheltered brackish water habitats in salt pans. Can be found in muddy ditches and salt-marsh areas near coast.

Distribution: Coastal areas with intertidal habitats and salt marshes.

Voice: Repeated thin twittering.

Status: Scarce Migrant.

Winter

Winter

Distinguished from Little Stint (see p. 121) by greenish-yellow legs, and grey-brown head and breast (white in Little). Breast-band sharply demarcated from white belly. Sides of tail white in Temminck's, grey in Little. Back plain and lacks white mantle stripes of Little.

Winter

Summer

LONG-TOED STINT
Calidris subminuta

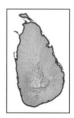

Size: 14cm

Habitat: Favours freshwater habitats such as marshes and even fallow paddy fields near coast.

Distribution: Close to coast, in areas favoured by other shorebirds.

Voice: Repeated *teeuw* that may vary in pitch and pace. Flight call is more trilling.

Status: Scarce Migrant.

Long tibia. Pale base to bottom mandible evident even in juvenile. In winter, mantle feathers grey with dark centres. In summer, mantle feathers tinged russet, with dark centres with an elongated 'needle' and white tips. Summer Rufous-necked Stint (see p. 121) has a similar pattern, but mantle is more rufous and neck is also rufous.

SHARP-TAILED SANDPIPER
Calidris acuminata

Size: 19cm

Habitat: On breeding grounds found in freshwater wetlands. In wintering areas found on mudflats and in other brackish habitats.

Distribution: Coastal areas; most records are from dry lowlands.

Voice: Rapidly uttered sequence of *whi-whi* notes with changes in pace and pitch.

Status: Vagrant.

Rufous-tinged crown. Orange-buff breast not as sharply demarcated from white belly as in Pectoral Sandpiper (see right). Streaking mainly on sides of neck. Thick white eyebrow and white eye-ring. Facial pattern stronger than in Pectoral. Flanks have V-shaped markings.

Winter

Winter

Winter

PECTORAL SANDPIPER
Calidris melanotos

Size: 21cm

Habitat: Favours freshwater wetlands. Can turn up on mudflats.

Distribution: May turn up anywhere in lowlands on coastal intertidal areas or freshwater wetlands.

Voice: Flight call a repeated and tremulous *kreet*.

Status: Vagrant.

Breast streaks end abruptly. Streaked breast demarcated sharply from white belly in all plumages. Downcurved, blunt-tipped bill. Long primaries (longer than tertials) extend to tip of tail, giving it an attenuated look side-on despite it being a bulky and broad-chested bird. Broad dark line across rump.

Winter

DUNLIN *Calidris alpina*

Size: 20cm

Habitat: Mudflats and intertidal habitats on coasts and wide rivers.

Distribution: Likely to turn up on coastal mudflats. Most records are from northern half of Sri Lanka.

Voice: Contact call a repeated, tremulous, short *kreet*. Song a long, undulating trill with variations.

Status: Vagrant.

Winter

Winter

Very similar to common Curlew Sandpiper (see right). Latter has evenly downcurved bill; in Dunlin downcurve is more like a slight kink towards bill tip. Diagnostic characteristic is that in flight, a longitudinal dark stripe is seen to run through middle of rump, absent in Curlew Sandpiper. Winter Dunlin can have fine streaking on breast, extending to sides. Summer Dunlin is distinct, with black belly-patch (absent in Curlew Sandpiper) and chestnut-and-black patterned scapulars.

CURLEW SANDPIPER *Calidris ferruginea*

Winter

Size: 21cm

Habitat: Estuaries and mudflats in dry lowlands.

Distribution: Throughout Sri Lanka on estuaries, salt pans and other coastal habitats. Bundala and Palatupana Salt Pans are reliable sites for seeing flocks at close range.

Voice: Rapid, tremulous, twittering call. Background call heard most frequently in estuaries in Sri Lanka where mixed wader flocks are found.

Status: Common Migrant.

Distinctive small wader with downcurved, long bill. In winter plumage, which it wears most of the time in Sri Lanka, has white underparts with grey upperparts. Mantle and flight feathers are grey with a thin blackish shaft-streak. Longer legs than those of Dunlin (see left), and slightly different bill shape. In breeding plumage, which some birds acquire before they leave Sri Lanka, head, neck and breast turn rufous. Mantle feathers have rusty base with black 'trident' marking. Thick dark line bifurcating white rump in Dunlin. Curlew Sandpiper has unbroken white rump. Both species look very similar in winter, and both have thin white wing-bars. In summer plumage, Dunlin acquires black belly and lacks rufous underparts and head of Curlew Sandpiper.

Spoon-billed Sandpiper

Eurynorhynchus pygmeus

Winter

Summer

Size: 15cm

Habitat: Intertidal coastal mudflats.

Distribution: May occur on coast. Mudflats in Mannar area may be the best places to locate this vagrant.

Voice: On breeding grounds, vocalizations include short, shrill buzzes, *zhree-zhree-zhreee*.

Status: Vagrant.

Large-headed, stint-like bird with conspicuous, large spatulate bill. 'Spoon-bill' is most evident in a head-on view. In breeding plumage, head turns rufous. In winter plumage, similar to Rufous-necked Stint (see p. 121), but has more white on forehead and a darker and more spotted cap.

Buff-breasted Sandpiper

Tryngites subruficollis

Size: 19cm

Habitat: Often in pastures and fields with short grass. In its breeding range may be seen on golf courses. Also intertidal mudflats.

Distribution: Mudflats and short grass pastures beside coast.

Voice: Contact call a sharp *chik* or *chuk* repeated a few times.

Status: Vagrant.

Squarish head and straight black bill give this species a distinctive profile. Overall buff. Juvenile has white fringes to feathers. Adults have warmer edges and more loosely arranged mantle feathers. Attenuated with long wings extending to tail. Long yellow legs. Lacks streaked bib of Pectoral Sandpiper (see p. 123).

Winter

Winter

BROAD-BILLED SANDPIPER
Limicola falcinellus falcinellus

Size: 17cm

Habitat: Lagoons, salt pans and similar areas on coastline.

Distribution: Coastal habitats in dry lowlands where very shallow water levels are found. Due to its short legs, feeds in shallow water. Seems to be a rather wary sandpiper (I have never seen it feed close to a roadside even in salt pans, where other waders are tolerant of vehicles).

Voice: Rapid, tremulous chirruping notes, reminiscent of a bunting or other grassland passerine.

Status: Uncommon Migrant. Scarce but regular visitor.

Winter

Similar shape to Curlew Sandpiper (see p. 124), small size and active habits help to locate species in a mixed flock of waders. Look out for 'stitching action' when the birds feed. Split supercilium not always apparent, even with good views.

Male, Winter

Female, Winter

RUFF Philomachus pugnax

Size: m-31cm/f-25cm

Habitat: Mudflats, estuaries and lagoons in lowlands.

Distribution: All around lowland coasts where suitable habitat is found. Salt pans are a good place to look for it. May occasionally be seen in freshwater habitats such as paddy fields in transit with other waders.

Status: Uncommon Migrant.

Plump-bodied, small-headed appearance gives this wader a profile that helps to pick it out from a flock. Also bigger than most *Calidris* waders, it is usually in mixed species concentrations. Mantle and wing feathers have pale edges, giving it a scalloped effect. Adult females and juveniles have all-dark bills. Adult female usually has brighter coloured yellowish legs than juvenile. Males have pale base to bills and are larger than females.

> **Stilts and avocets** Family *Recurvirostridae*
> Stilts are distinctive waders with long bills, long
> pink legs and black upperparts. The Avocet
> is a distinctive black-and-white wader with a
> downcurved bill.

BLACK-WINGED STILT
Himantopus himantopus himantopus

Size: 25cm

Habitat: Fresh and brackish waterbodies in lowlands.

Distribution: During the migrant season there is an influx of birds to wet zone, but it is almost entirely absent from wet zone outside this season. It is highly likely that birds resident in dry zone are joined by migrants from the Asian mainland. Most birds in wet zone will be these birds. However, note that in April 2010, there was a second record of these birds breeding in wet zone in Talangama. Possibly half a dozen pairs had attempted to nest and at least two pairs raised young. In winter, some birds are seen with a black hindneck, reminiscent of Australian Stilt, which is another subspecies. It is likely that these birds are also migrants from the Asian mainland.

Voice: Shrill, insistent *ack ack*.

Status: Common Resident and Migrant.

Male

Female

Females have browner backs than males. Juveniles are marked with grey on face and hindneck.

PIED AVOCET
Recurvirostra avosetta

Size: 46cm

Habitat: Lagoons and estuaries. Often wades into water to feed.

Distribution: Scarce migrant mainly to northern half of Sri Lanka. In some years, avocets are recorded from Mannar area in numbers varying from 30 to 175. In other years there are no records at all anywhere in Sri Lanka.

Voice: A *whee* is repeated with some spacing between the single note.

Status: Scarce Migrant.

Thin, upturned black bill and black-and-white plumage make this wader distinctive. In flight shows a black wing-tip and two black bands on white inner wing. Long greyish legs.

Phalaropes Family *Phalaropodidae*
These waders have a lot of buoyancy and feed off the water's surface. They are scarce migrants to Sri Lanka. Because of their buoyancy in water, they are famous for turning up as vagrants far from their usual wintering grounds in other parts of the world.

Crab-plover Family *Dromadidae*
This is a black-and-white wader with a stout, conical bill. It is found along coastlines on the Indian Ocean from Africa to Asia. There is only one species in the family.

RED-NECKED PHALAROPE
Phalaropus lobatus

Size: 19cm

Habitat: Salt pans and estuaries in dry lowlands.

Distribution: Bundala National Park is one of the best sites.

Voice: Shrill, sharp *clip clip*. Can sound a little tremulous.

Status: Scarce Migrant.

In breeding plumage, develops a red neck, white throat, black head and rusty edges to mantle feathers. Nearly parallel lines on back change from white in winter to rust in summer. In winter plumage, in which it is usually seen in Sri Lanka, it looks a black-and-white bird with a black smudged eye-patch behind the eye. Thin black bill with just a hint of being upturned is a key feature. In flight shows a white wing-bar. Juveniles have a dark crown and browner upperparts. Often seen swimming in the water and not 'wading'. The only phalarope species accepted as being reliably recorded up to now in Sri Lanka. Red Phalarope P. fulicarius is recorded as a vagrant in northern India. In winter plumage, adult Red Phalarope's back is plain and lacks white stripes on mantle and scapulars of wintering adult Red-necked.

Winter

CRAB-PLOVER *Dromas ardeola*

Size: 41cm

Habitat: Mudflats in estuaries, tidal lagoons and other such areas, where it can hunt for marine and other tide-line invertebrates including crabs.

Distribution: Scarce breeding resident in North-west Sri Lanka in strip to north from Mannar. Birds may take part in local migrations and have been seen on south-east coasts. It cannot be ruled out that some birds may be migrants from mainland Asia.

Voice: Three-noted call that sounds like *ka-hee-kew*. Distinct call that cuts through background noise. Some calls are reminiscent of a puppy yelping.

Status: Scarce Resident.

Black-and-white, stoutly built wader with a conical, thick black bill. Pied Avocet (see p. 127), another black-and-white wader, has a thin, upturned bill and is longer necked with a black crown. Juvenile duller with a greyish or dirty-white mantle. In flight, black flight feathers contrast with white forewing. The disruptive colouration against a dull greyish-brown mudflat can at times make it surprisingly hard to pick out these birds.

> **Stone-curlews and thick-knees** Family
> *Burhinidae*
> These nocturnal birds with cryptic camouflage have thick-set bills and large eyes for night vision.

INDIAN STONE-CURLEW
Burhinus indicus

Size: <41cm

Habitat: Grassland in lowlands, often where there is adjoining scrub forest.

Distribution: Mainly in dry zone. Often overlooked in wet zone due to nocturnal habits. Can often be heard from golf course in central Colombo,

Voice: Utters a series of ascending *yip yip* notes when calling. Notes change in pitch suddenly.

Status: Scarce Resident. Seems to be more scarce than Great Thick-knee.

Relatively large eyes. Can be distinguished from Great Thick-knee (see right) by its strongly streaked body. Furthermore, unlike in Great Thick-knee, bill is short, stubby and straight. Great has slightly upcurved bill.

GREAT THICK-KNEE
Esacus recurvirostris

Size: 51cm

Habitat: Mainly in dry lowlands in grassland.

Distribution: Appears to be confined to dry lowlands close to coast. Feeds on invertebrates on grassland. It is not clear why it prefers grassland habitats with adjoining scrub forest close to coast. National parks such as Yala and Bundala are good sites to see this bird.

Voice: Utters a melancholy *till-leowp*. Often heard at night and is a signature of the dry-zone forests' nightscape.

Status: Uncommon Resident.

Has large eyes and is mainly nocturnal. Scavenges on carcasses, but it is not certain if it eats meat off carcasses or just pulls out invertebrates feeding on a carcass. Confusion is possible with Indian Stone-curlew (see left), but Great Thick-knee has more uniform upperparts and long, upcurved bill.

> **Coursers and pratincoles** Family
> *Glareolidae*
> Coursers are ground-dwelling birds that have a
> hurried gait when feeding. Pratincoles are long-
> winged waders.

INDIAN COURSER
Cursorius coromandelicus

Size: 26cm

Habitat: Dry lowland grassland, close to coast. Seems to be happy in grass taller than that frequented by birds such as lapwings. When feeding, runs in little spurts in a plover-like fashion.

Distribution: Recently only known from a few sites in Mannar district, and from Jaffna Peninsula and Delft Island. Occurs in small flocks of around half a dozen individuals in known sites. One of the rarest residents.

Status: Highly Scarce Resident. Critically Endangered on IUCN Red List.

Coursers are like plovers with shortish, downcurved and pointed bills. Indian Courser is the only courser found in Sri Lanka. Adults have a chestnut crown and neck. Face has a black line from base of bill to nape, with a white line over it. Juvenile has the stripes, but in shades of brown and white. In juvenile, mantle feathers are edged with white.

COLLARED PRATINCOLE
Glareola pratincola

Size: 23cm

Habitat: Dry, flat areas with a scattering of short grassland. Also in salt pans and other dry areas close to coastal habitats. Usually seen on dry land, unlike other waders that prefer wet or muddy habitats.

Distribution: Coastal areas. Most records are from South-east near Bundala National Park.

Voice: Well-articulated, nasal *whi-whi*.

Status: Scarce Migrant.

Similar to Oriental Pratincole (see p. 131). Key diagnostic feature is thin white trailing edge on wings that is absent in Oriental. Forked tail longer than in Oriental. Usually at rest, tail is longer than folded wing-tips. Upperwing-coverts grey-brown and contrast with black outerwing. In Oriental the contrast is less as wing-coverts are darker.

Breeding

ORIENTAL PRATINCOLE
Glareola maldivarum

Size: 24cm

Habitat: Similar to Collared Pratincole, preferring dry flatlands.

Distribution: Mainly in coastal dry lowlands from Northern Peninsula to South.

Voice: Clear-sounding, repeated *ti-ti-treeup*. Tremulous at the end.

Status: Uncommon Resident.

Similar to Collared Pratincole (see p. 130), but distinguished from it by lack of white trailing edge, darker wing-coverts and shorter forked tail (see also under Collared Pratincole). In breeding plumage, Oriental and Collared Pratincoles acquire black gorget and have red bill base.

Non-breeding

Breeding

SMALL PRATINCOLE
Glareola lactea

Size: 17cm

Habitat: In open areas adjoining waterbodies in dry lowlands.

Distribution: Dry lowlands close to coast. Often seen in small flocks.

Voice: Metallic grating note repeated regularly.

Status: Uncommon Resident. Vulnerable on IUCN Red List.

Easily told apart in flight from similar Oriental and Collared Pratincoles (see left and p. 130) by conspicuous black-and-white wing-bars on flight feathers. Inner primaries show some white on trailing edges. In breeding plumage has a black line from bill to eye. Tail only weakly forked. Has black underwing-coverts that show in flight.

Telling Apart the Pratincoles

	Small	Oriental	Collared
Wing-bar	Prominent, black-and-white bars on flight feathers	NA	NA
Upper forewing	Light sandy-brown	Darker.	Dark brown, little contrast with black primaries
Trailing edge of secondaries	Thick black bar	NA	Thin white edge
Throat gorget	No	Yes	Yes

Skuas Family *Stercorariidae*
Skuas are pelagic birds that engage in kleptoparasitism. In Sri Lanka, they are usually far out at sea unless blown towards the land by storms. In North America they are known as jaegers, and in Europe as skuas. Females are bigger than males, a feature shared with birds of prey (raptors and owls) and frigatebirds, where large prey is taken, or kleptoparasitism (frigatebirds) is a key feeding strategy. Kleptoparasitism is most developed in skuas, with the Arctic Skua taking almost all of its food over the sea by stealing it from other seabirds like terns.

Skuas can be difficult to identify unless they are adults with the tail showing the typical pattern without any damage or wear distorting its shape. Advanced guides need to be consulted to identify immatures. Skuas are seen just before the south-west monsoon strikes Sri Lanka. They are also seen during the mass migration of Brown-winged Terns in August–September, when they are travelling south past the west coast.

BROWN SKUA *Catharacta antarctica*

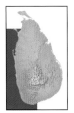

Size: 58cm

Habitat: Pelagic; passage migrant.

Distribution: Most records are from the west coast.

Status: Scarce Migrant.

Larger than South Polar Skua (see p. 133), though size may not be apparent in the field. Best field identification characteristic is identical colour of head, body and wings. Underwing pattern of dark wing lining, paler underwing-coverts and pale patch at base of primaries. South Polar Skua has similar pattern.

SOUTH POLAR SKUA
Catharacta maccormicki

Size: 53cm

Habitat: Pelagic; passage migrant.

Distribution: Most likely to be seen in the seabird passage on west coast.

Status: Vagrant.

Juvenile

Smaller than Brown Skua (see p. 132), though size may not be apparent in the field. Best field identification characteristic is contrast between paler head and darker body and wings. Underwing pattern of dark wing lining, paler underwing-coverts and pale patch at base of primaries. Brown Skua has similar pattern.

POMARINE SKUA
Stercorarius pomarinus

Size: 56cm

Habitat: Pelagic. Passage migrant seen during mid- to late April before south-west monsoon.

Distribution: Flocks of up to 20 birds have been seen on sailings from Mirissa, though single birds are more likely to be seen. May also (rarely) be seen off Kalpitiya.

Status: Migrant.

Typical adult has twisted tail-tips, giving a rounded, blunt tip. More pot-bellied than Parasitic (Arctic) Skua (see p. 134).

PARASITIC (ARCTIC) SKUA
Stercorarius parasiticus

Size: 45cm

Habitat: Pelagic. Passage migrants seen during mid- to late April before south-west monsoon.

Distribution: Most records are from sea watchers and whale watchers off the west coast.

Voice: Squawking calls; high-pitched mewing call, *kyaa-kyaa-kyaa*.

Status: Highly Scarce Migrant.

Adult, dark form

Has pointed tail-tips, though Pomarine Skuas (see p. 133) with abraded tail-tips could be mistaken for Parasitic Skuas. Narrower wing-bases than in Pomarine, which has broad-based wings contributing to its chunky look.

LONG-TAILED SKUA
Stercorarius longicaudus

Size: 41cm w/o streamers

Habitat: Pelagic. Passage migrant seen in mid- to late April before south-west monsoon.

Distribution: The observation referred to below may be only the second or third record in the South Indian region. Probably overlooked due to a lack of observers sea watching.

Status: Vagrant.

Long tail-streamers are distinctive (Riaz Cader and I photographed a bird sitting on the sea off Kalpitiya on 11 April 2010 – it reminded me of a Pheasant-tailed Jacana, see p. 101, because the tail-streamers were so long). Adults have clean, dark caps. Upperwing brown-grey, lighter than in other skua species. Juvenile dark with barred undertail-coverts, and lacks long tail-streamers.

Gulls Family *Laridae*

Gulls are short-legged birds that spend a lot of time in the air looking for food to pick off the surface of the water. Large gulls even take small mammals.

This is a group that fascinates birders because of the challenges posed by ageing and identifying immatures as they change into full adults. In Sri Lanka, juveniles of the two large gulls, Pallas's and Heuglin's Gulls, are usually seen with adults, which makes it easier to figure out which species they belong to. Gulls seen in Sri Lanka have travelled great distances to winter there, with some having journeyed from the Russian Arctic.

Moult and Ageing in Gulls

All gulls have complex plumages with a fairly well-defined pattern of change that allows birders to 'age' them. Ageing and identifying difficult gulls is a technical challenge relished by advanced birders.

No gulls are resident in Sri Lanka. They are all winter visitors. The gulls belong to three age groups, as follows, with birds increasing in body size falling into longer age groups.

Two-age groups Smaller gulls such as Brown-headed Gull, a regular migrant to Mannar.

Three-age groups Species such as Common Gull (not recorded in Sri Lanka).

Four-age groups Species such as Heuglin's Gull.

On their breeding grounds in summer, each of these age groups has an additional group comprising recently fledged juveniles. In Sri Lanka all birds are at least in first-winter plumage.

All birds change their plumage in a succession of moults. The first moult, or post-juvenile moult, takes place in autumn and involves just a change in the head and body feathers. The spring moults are of the head and body feathers. The spring moult or partial moult may take 1–2 months to complete. Therefore, a bird in its first summer has undergone a post-juvenile moult and a spring moult, both of which have involved only the head and body feathers. Its tail and wing feathers will be rather worn and ragged. In successive autumnal moults all of its feathers are replaced. The autumn moult is a complete moult and may take 3–4 months.

Juvenile gulls have dark bands on their tails and brownish feathers on their wings, which they gradually lose over a succession of winter plumages. In the sequence below of Heuglin's Gulls, note how the tail band and brown wing feathers progressively diminish with age.

1st Winter *(August)*

2nd Winter *(August)*

3rd Winter *(January)*

Adult Winter *(January)*

Urumalai Beach in Mannar provides bird photographers with wonderful opportunities for photographing gull plumages, and of course other coastal birds like waders and egrets.

SOOTY GULL Larus hemprichii

Size: 45cm

Habitat: Bird of the coastline, not shy of human company in fishing ports.

Distribution: Coastline; a few records from South, but may turn up anywhere on coastline, especially in Northern Peninsula and Mannar.

Voice: Melancholy, hollow-sounding *ow*, repeated at intervals.

Status: Vagrant.

Summer

Immature

Head looks disproportionately small to amount of bill, giving the birds a distinctive small-headed look. Adult has dark head, white collar around nape, and dark upperwing and underwing. Yellowish bill has a two-tone tip of black and red. First winter is grey-brown and has a broad, dark terminal band on tail.

2nd Winter

Adult

HEUGLIN'S GULL Larus heuglini

Size: 60cm

Habitat: Mainly in coastal areas.

Distribution: Northern parts of Sri Lanka.

Voice: Nasal, drawn-out call notes. Very similar to Lesser Black-backed Gull, common in Europe.

Status: Migrant.

Treated by some as a race of *L. fuscus*. Breeding adult has dark grey upperparts with a distinctive white trailing edge to wings. Yellow-legged Gull *L. michaehellis* has paler upperparts and different pattern on wing-tip, with more white on tips of two outermost primaries (P9 and P10). Juveniles have dark tail-tips and dark brown flight feathers. The taxonomy of white-headed gulls remains in flux, and Steppe Gull (see p. 137) is currently treated as a subspecies of Heuglin's Gull. Steppe has paler back and white spot on outer two primaries. Juvenile is whiter on head and underparts than Heuglin's. Steppe Gull is a vagrant to Sri Lanka and may be overlooked. In Heuglin's as it matures, red spot on yellow bill turns to red-and-black spot.

STEPPE GULL Larus [heuglini] barabensis

Size: 60cm

Habitat: Winters on coastlines.

Distribution: Coastal areas in Northern Peninsula where gull flocks are regularly encountered are the most likely to include this vagrant.

Voice: Similar to Heuglin's Gull.

Status: Vagrant.

Summer

Summer

The best field identification characteristic is a smallish dark eye, giving this species a placid look. Heuglin's Gull (see p. 136) has a pale eye and fierce look. Dark grey back, but paler than in Heuglin's. Similar Caspian Gull L. *cachinnans* (not recorded in Sri Lanka) has long and parallel-sided bill, and is long-legged with an upright stance. Eye in Caspian Gull is a dark 'bullet-hole'. Steppe Gull has at least a hint of pale irides (sometimes more prominently pale irides), and is more round headed. Caspian has pale grey back.

Summer

Winter

GREAT BLACK-HEADED (PALLAS'S) GULL Larus ichthyaetus

Size: 69cm

Habitat: Seagoing bird. During the day, birds often roost in open fields.

Distribution: Most common in Northern Peninsula. Kora Kulam in Mannar is a reliable site for roosting gulls.

Voice: Deep, murmuring call with a melancholy accent.

Status: Migrant.

Large size and tricoloured bill. Juveniles have grey on mantle. Lacks 'white mirror' on dark primary tips found on Brown-headed Gull (see p. 138). Has dark iris.

BROWN-HEADED GULL
Chroicocephalus brunnicephalus

Size: 42cm

Habitat: Coastal regions in relatively unpolluted or undisturbed areas.

Distribution: Around coastline, concentrated in areas where fish are brought ashore. Numbers higher in northern half of Sri Lanka.

Voice: Strident, mewing call.

Status: Migrant.

Winter (Adult)

Summer

Similar to Common Black-headed Gull (see right). Brown-headed has 'wing-tips dipped in black ink', with a white mirror between point of wing-tip and concave base of black triangle. Pale iridies (Great Black-headed, p. 137, has dark iridies), and slightly more distinct pale eye-ring than in Great Black-headed, give it a different look. In breeding plumage head turns chocolate-brown. In winter white head has smudgy 'headphones' pattern on head. First winter has black terminal band on tail and banding on wing-coverts.

COMMON BLACK-HEADED GULL
Chroicocephalus ridibundus

Winter

Summer

Size: 38cm

Habitat: Coasts. In Europe common on inland freshwater bodies.

Distribution: Most records are from North-west coastline.

Voice: Grating, loud *kraa*.

Status: Scarce Migrant.

In adult, clean white wedge on leading edge of wing, with rear of white wedge lined in black. White wedge strongly contrasts with grey upperwing. Tips of primaries lack white mirrors of Brown-headed Gull (see left). When viewed from rear, even at a distance, leading white edge on forewing and white tail can be seen to contrast with grey upperwing. In winter, white head has smudgy 'headphones' pattern on head. First winter has black terminal band on tail and banding on wing-coverts.

SLENDER-BILLED GULL *Larus genei*

Size: 42cm

Habitat: Bird of coastal waters.

Distribution: May occur anywhere on coast. Most records in Sri Lanka are from coastal strip in northern half.

Voice: Throaty *kraa-kraa*.

Status: Vagrant.

Slightly longish, thin bill and shallowly sloping forehead give this gull a distinctive thin, long-headed look. Red bill and orange legs. White head has faint grey patch on ear-coverts in winter plumage. Pale irides in adult. In juvenile, eye and bill are dark.

Adult (left), Juvenile (right)

Adult

> **Terns Family *Sternidae***
> Slenderly built, with long wings, and often with long outer tail feathers. Most terns pick food off the water's surface; a few plunge dive.

Summer

Winter

GULL-BILLED TERN

Gelochelidon nilotica nilotica

Size: 38cm

Habitat: Widespread visitor to waterbodies in lowlands. Most common along coastal areas. On arrival spreads widely in coastal areas.

Distribution: Mainly found on coastline patrolling beaches.

Voice: Double-noted *cheez-weet*. Nasal, and cuts through the air.

Status: Scarce Resident and Common Migrant.

Heavy dark bill. Breeding adult has black cap. In non-breeding plumage, eyes have dark smudge extending behind them. Crown has traces of black cap.

CASPIAN TERN Hydroprogne caspia

Size: 51 cm

Habitat: Bird of the coast, salt pans and estuaries. Occasionally may be seen hunting over freshwater bodies not far from coast.

Distribution: Mainly in coastal areas. Congregates in sites such as salt pans.

Voice: Drawn-out, screechy whistle. Also some harsh, grating calls.

Status: Scarce Resident and Migrant.

Large tern with sturdy red bill and legs. Black crown in summer; in winter, forehead is white with traces of black. Leading primaries are black or dusky on underwing, and upperwing is tipped with black. Black legs. Unlikely to be confused with other terns in Sri Lanka. Juvenile has scaly mantle with greyish feathers edged with brown. Black crown is underdeveloped.

Winter

Winter

LESSER CRESTED TERN
Thalasseus bengalensis bengalensis

Winter

Winter

Size: 43cm

Habitat: Seagoing tern that can be seen roosting in brackish water habitats (salt pans, estuaries and similar places).

Distribution: All around coastal areas.

Voice: Grating, tremulous *thureep* and variations on this theme.

Status: Migrant.

Orangeish bill (Great Crested Tern, see p. 141, has yellow bill).

GREAT (LARGE) CRESTED TERN
Thalasseus bergii

Size: 47cm

Habitat: Seagoing tern that can be seen roosting in brackish water habitats (salt pans, estuaries and similar places).

Distribution: All around coastal areas.

Voice: Harsh, far-carrying *kraa*.

Status: Race *velox* is a Resident and race *thalassina* is a Vagrant.

Greenish-yellow bill. Shallow and measured wingbeats. Viewed directly from below, looks as though it is hardly beating its wings at all. Engages in plunge diving to catch fish.

Winter

Winter

Winter

Summer

SANDWICH TERN *Thalasseus sandvicensis*

Size: 43cm

Habitat: Hunts for fish over sea or waterbodies close to sea, and often with a connection to the sea. Breeds on isolated beaches.

Distribution: In Sri Lanka may turn up beside sea.

Voice: Sharp, grating *cheerweer*, repeated with spacing.

Status: Highly Scarce Migrant.

In breeding plumage distinctive, with yellow-tipped, long black bill and shaggy black crest. Non-breeding birds lack yellow tip, but have traces of black crest on hindcrown as well as flat heads. Head shape different from that of *Sterna* terns, which have rounded heads. Short-legged tern.

ROSEATE TERN *Sterna dougallii*

Size: 38cm

Habitat: Pelagic. May turn up on beaches.

Distribution: Considered a scarce breeder on rocky islets around coast and on Adam's Bridge islands in North-west. Breeding is in April– August; rarely seen outside these periods.

Voice: Sharp, short, repeated *krik*. Nasal.

Status: Uncommon Resident.

In breeding plumage similar to Common Tern (see p. 143). Bill can be black-tipped or entirely coral red in local race. Breeding birds develop rosy tinge to underparts. Long white tail-streamers. Upperparts lighter and underparts whiter than in Common. Non-breeding plumage similar to Common's, and calls are best means of distinguishing the species.

BLACK-NAPED TERN
Sterna sumatrana

Summer

Summer

Size: 33cm

Habitat: Inshore waters around lagoons.

Distribution: Pelagic. Breeding visitor to Andamans and Maldives, with vagrants to Sri Lanka that may potentially turn up anywhere at sea.

Voice: Sharp, grating *treep*, repeated with spacing.

Status: Vagrant.

Long black bill. Black line through eye continues to nape, forming band from eye to eye looping behind neck. White elsewhere. In non-breeding plumage nape-band is less distinct.

COMMON TERN *Sterna hirundo*

Size: 36cm

Habitat: Coastal areas; roosts in salt pans, estuaries and similar places.

Distribution: Widespread on arrival in coastal areas. Has been been overlooked in the past because many observers were unable to identify the species. They are regular visitors to the strip from Chilaw Sand Spit to Mannar.

Voice: Loud, grating *kraa* calls alternating with high-pitched, shrieking *kree* notes.

Status: Most birds are Migrants; also believed to be Scarce Resident.

Winter

Summer

Juvenile and winter birds show dark carpal bar in flight and at rest. Bill longer than in *Childonias* marsh terns. In the hand or on close views, bill looks stouter. In summer, bill and legs are red. In winter, legs can turn black, and in some individuals retain varying degrees of redness. Outer primaries dull black. Mantle and upperparts grey. Black cap has receded, exposing white forehead. Birds seen are believed to belong to race *longipennis*. It is possible that race *tibetana* also occurs.

Winter

LITTLE TERN *Sternula albifrons sinensis*

Size: 23cm

Habitat: Frequents both the coast and inland freshwater lakes in lowlands.

Distribution: Breeding resident in dry zone and visitor to wet zone. Breeding birds may be supplemented by wintering birds from Asian mainland, but this is to be proven.

Voice: Repeated, high-pitched *chick chick*.

Status: Resident.

Smallest of terns, with pointed wings and rapid, butterfly-like, fluttering flight. In breeding plumage, develops black crown with white forehead, and black line from base of bill through eye to back of black crown. Yellow bill with black tip. In non-breeding birds, bill is all black but may retain yellow base. Black crown recedes further from white forehead and is more diffused and duller. Scarce Saunders's Tern (see p. 144) has squarer white forehead without hint of white supercilium as seen in breeding plumage of Little Tern.

SAUNDERS'S TERN *Sterna saundersi*

Size: 23cm

Habitat: Bird of coasts rarely straying inland, unlike similar Little Tern, which can at times appear at freshwater lakes inland. In breeding range breeds on sandy islands and isolated beaches.

Distribution: May show up anywhere on coastline. Breeding suspected on Adam's Bridge Islands, but no recent conclusive evidence.

Status: Highly Scarce Resident.

In breeding plumage, told apart from Little Tern (see p. 143) by 'square-cut' white forehead. In Little Tern white on forehead extends behind eye. Saunders's also has darker greyish rump and tail. Black primary wedge on outer wing more prominent in Saunders's. In non-breeding plumage, there are no reliable field characteristics that can be used to separate the two species.

Winter

Summer

Summer

Summer

WHITE-CHEEKED TERN *Sterna repressa*

Size: 35cm

Habitat: Coastal and pelagic.

Distribution: Sight records off the west coast and ringing records from the South-east in Bundala. May turn up anywhere on coast.

Voice: Repeated, drawn-out, grating *cheer.*

Status: Vagrant.

Told apart from similar Common Tern (see p. 143) by uniformly grey rump, and uppertail-coverts that are concolorous with the back. Adult has long tail-streamers. In non breeding plumage, thick black line from eye to nape. Black bill. Juvenile dirty grey on upperwing with dark leading edge to innerwing.

BROWN-WINGED (BRIDLED) TERN
Sterna anaethetus antarctica?

Size: 37cm

Habitat: Pelagic. Large numbers pass along the west coast in August–September.

Distribution: One of the most frequently seen seabirds on whale watches, but seen by only a tiny handful of shore-based birdwatchers before May 2008.

Status: Common Migrant and Highly Scarce Resident.

Uniform grey-brown upperparts in adult and pale under-body. Tail brownish. Can look black under certain lighting conditions. In adult, upper tail and uppertail-coverts are brown. Underwing-coverts are white with brown flight feathers. Wing-coverts and mantle slightly paler brown than primaries, but this is not obvious in the field. White extends over eye, forming a short but thick supercilium. In juveniles, uppertail-coverts can be pale. Bill looks long and thin compared with bills of other terns. Adult has elongated outer tail feathers; the 'tail-streamers'.

Breeding adult

Immature

Breeding adults

SOOTY TERN *Sterna fuscata nubilosa*

Size: 43cm

Habitat: Pelagic bird, rarely seen near shore unless stormy weather brings it close.

Distribution: Sooty Terns have been seen breeding on islands on Adam's Bridge, but the majority of birds appear to pass through on passage. Kalpitiya seems to be one of the best places to see it, with reports of exhausted birds seen resting on the beach. I photographed an immature off Kandakkuliya on the Kalpitiya Peninsula.

Status: Mainly Scarce Migrant and Highly Scarce Resident.

Adult has black upperparts. Similar Brown-winged (Bridled) Tern (see left) has brown upperparts, but in awkward lighting confusion is possible. A diagnostic feature is that white supercilium does not extend behind eye in Sooty Tern. Juvenile Sooty is heavily marked on mantle and inner wing with white bars formed by pale tips to feathers.

WHISKERED TERN
Chlidonias hybrida hybrida

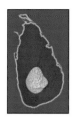

Size: 25cm

Habitat: Bird of freshwater lakes and marshes with open stretches of water. May occasionally be seen on beaches, especially before return migration.

Distribution: On arrival spreads up to mid-hills.

Voice: Repeated metallic *krick*.

Status: Common Migrant.

Summer

Winter

In breeding plumage, reddish bill and black crown contrast with white cheeks. Reddish legs. Belly turns black, contrasting with whitish underwing. Undertail white. In non-breeding plumage, bill and legs are black, forehead is white, and there is a dark smudge over eye and traces of black on rear end of crown towards nape. Juveniles are like non-breeding adults with dark mantles scalloped with white edges to mantle feathers.

WHITE-WINGED TERN
Chlidonias leucopterus

Winter

Summer

Size: 23cm

Habitat: Freshwater lakes, lagoons and similar places in lowlands.

Distribution: On arrival spreads around lowlands, not ascending as high as Whiskered Tern. Seems to favour dry lowlands.

Voice: A *krick* call, more higher pitched than that of Whiskered Tern.

Status: Migrant.

In all plumages, White-winged Tern has a contrasting white tail and rump. In breeding plumage, black underwing easily separates it from Whiskered Tern (see left). Non-breeding birds can be confused with Black Tern C. niger, which may be a vagrant to South Asia. Black has more pronounced forked tail and grey rump (white in White-winged). Non-breeding Black has greyish-brown side-patches, and broader black patch on cheek linked to black patch on hindcrown. Breeding birds have smoky black upperparts, unlike strongly contrasting white wings in breeding White-winged.

Brown Noddy *Anous stolidus pileatus*

Size: 41cm

Habitat: Pelagic.

Distribution: Pelagic; passage visitor just before south-west monsoon strikes Sri Lanka. Seas off Kalpitiya seem to be the best location. Also seen occasionally by whale watchers from boats off Mirissa.

Voice: Grunting note, repeated.

Status: Uncommon Migrant and Scarce Resident. Was considered an Uncommon Migrant until September 2014, when birds were seen breeding in Adam's Bridge Islands off Mannar.

Pale bar across upperwing-coverts contrasts with flight feathers. Underwing-coverts are paler and contrast with dark flight feathers, but this is not apparent unless lighting conditions permit it. Chocolate-brown tail contrasts with paler back. Pale forehead extends to crown. Can be very white in some birds. Lores dark, looking black. Bill more downcurved than Lesser Noddy's (see right). Bill heavy and jagged compared with Lesser Noddy's. Wing-flaps in flight distinctly slower and more laboured than those of Lesser Noddy. Tail also bigger and more spoon shaped compared with Lesser's. Also has a habit of bringing the tail down at right angles to the body when it needs to break air speed.

Lesser Noddy
Anous tenuirostris tenuirostris

Size: 33cm

Habitat: Pelagic.

Distribution: Pelagic; passage visitor just before south-west monsoon strikes Sri Lanka. Seas off Kalpitiya seem to be the best location. Also seen occasionally by whale watchers from boats off Mirissa.

Status: Highly Scarce Migrant. Probably under recorded, like many pelagics.

Smaller than Brown Noddy (see left), with fine long bill. Greyish lores contrast with dark around eye. Lores can often look white and in some birds this extends to crown, nape and upper back. More white on forehead than shown in illustrations in many field guides. Underwing uniformly brown, lacking contrast of Brown Noddy. Wingbeats quicker than in Brown Noddy.

COLUMBIFORMES
This order has a single family with over 300 species worldwide.

Pigeons and doves Family *Columbidae*
Doves are more slender than the stouter bodied pigeons. Pigeons are strong fliers, engaging in long-distance seasonal movements within Sri Lanka.

ROCK PIGEON Columba livia

Size: 33cm

Habitat: Rocky islands and areas with rocky cliffs providing nest sites.

Distribution: Pigeon Island off Trincomalee is believed to hold one of the true wild colonies. Can at times be seen in the interior in some national parks with rocky outcrops.

Voice: Similar to familiar cooing call heard from feral pigeons.

Status: Uncommon Resident.

Easily confused with feral pigeons, many of which show characteristics of their wild ancestor. The following characteristics are typical of the wild Rock Pigeon: white back that may be concealed at rest, dark bill and dark collar that may have a purple gloss. Grey upperwing has thick black trailing edge and neat black wing-bar on secondaries. Underwing pale except for narrow blackish trailing edge. Grey on head much darker than grey on wings.

CEYLON WOODPIGEON
Columba torringtonii ℮

Size: 36cm

Habitat: Large, forested stretches in highlands. Seasonal movements to lower hills, descending as low as Sinharaja.

Distribution: Horton Plains National Park and botanical gardens in Hakgala are two of the most reliable sites for it. Main walking trail of Sinharaja from Kudawa side is seasonally good – when the birds are present, they are tolerant of observers.

Voice: Deep, throaty *whoo*, with an owl-like quality.

Status: Uncommon Endemic.

Bluish-grey body with black-and-white pattern on purplish hindneck. Distinctly larger than a feral pigeon.

PALE-CAPPED WOODPIGEON
Columba punicea

Size: 36cm

Habitat: Found in a mix of habitats from tropical and subtropical forest, to secondary forest.

Distribution: May turn up in well-wooded forest patches with fruiting trees.

Voice: Described as short, mewing *coo*.

Status: Vagrant.

Large pigeon with brick-red body and wings, and prominent pale cap. Flight feathers and tail are dark. Juveniles are browner than adults and lack pale cap.

ORIENTAL TURTLE-DOVE
Streptopelia orientalis

Size: 33cm

Habitat: Occurs in a range of habitats from mountains to farmland.

Distribution: Most recent records from Sri Lanka have been in dry lowland national parks where Spotted Dove is found in abundance, but there are also records from wet lowlands and higher hills.

Voice: Repeated *kooo-kooo-coo-coo*, very similar to Eurasian Collared-dove's.

Status: Highly Scarce Migrant.

Stockier than Spotted Dove (see p. 150), with dark-centred chestnut scapulars and mantle feathers. Rump dark blue-grey. Neck has small, black-and-white striped patch, less extensive than black-and-white spotting on Spotted Dove. Tail pattern differs, with Oriental having dark tail tipped with white in race *meena* (central Asia) and greyish tips in race *orientalis* (Siberia). Spotted Dove in flight has black inner edge and white outer edge to tail. European Turtle-dove *S. turtur* (not recorded in Sri Lanka) has white belly. In Oriental, in race *meena*, belly has buffish tinge with white undertail-coverts. In race *orientalis*, belly and vent are grey-brown.

SPOTTED DOVE
Streptopelia chinensis ceylonensis

Size: 30cm

Habitat: Tolerant of human presence.

Distribution: Widespread throughout Sri Lanka. One of the most common birds in dry lowlands. Since the late 1990s it has been spreading into wet zone. By 2000 it had colonized the capital city, Colombo. As a child I never saw this bird in the wet zone, but I now hear it cooing in the mornings in Colombo. Expansion in range may be related to climate change.

Voice: Throaty and tremulous *kuk kuk krooo*, repeated many times.

Status: Common Resident.

Bold hindneck marking that is spangled black and white. Mantle feathers have dark bases, and wing-coverts have dark shaft streaks towards tips. All of this creates mottled pattern on upperparts.

RED COLLARED-DOVE
Streptopelia tranquebarica

Size: 23cm

Habitat: Open country that is lightly wooded.

Distribution: Likely to turn up in dry lowlands. In the 19th century was found to breed in Jaffna Peninsula, and there are records of vagrants to east coast.

Voice: Throaty, double *kroo-kroo* repeated rapidly.

Status: Vagrant.

Male unlikely to be confused with any other species in Sri Lanka, with a grey head and hindneck, brick-coloured body and wings with a white vent. Both sexes have black half-collar. Female is brown and can look like a small version of Eurasian Collared-dove (see p. 151). Tail pattern is different, with Red Collared-dove having a blackish tail base with outer half white. In Eurasian Collared-dove the dark-and-white areas are not 'square cut', and the dark areas diffuse into the white outer edges.

EURASIAN COLLARED-DOVE
Streptopelia decaocto

Size: 32cm

Habitat: Mainly in northern dry lowlands.

Distribution: Northern parts of Sri Lanka. Southern limit was around Nawadamkulama Tank. However, in the last few years it has been seen at Embilikala in the South. Another of the Deccan plateau species with a restricted distribution in Sri Lanka. Mannar is a good site for this bird.

Voice: Archetypical *whoo whoo whooo* cooing call of a dove. Notes often sound like *cu-ckoo*.

Status: Uncommon Resident.

EMERALD DOVE
Chalcophaps indica robinsoni

Size: 27cm

Habitat: Occupies remnant patches of forest. Seems to need densely shaded forests and has all but disappeared from environs of Colombo, though it is occasionally seen in suburbs around Talangama Wetland.

Distribution: Found throughout Sri Lanka.

Voice: Repeated *oom oom*. Birders are always surprised that this booming call originates from a tiny pigeon.

Status: Resident.

A fast-flying dash of green across a road in a national park is often how the Emerald Dove is seen. In Sinharaja, where the birds are habituated, it is possible to see them foraging on the ground as this is a ground-feeding dove. Small size and emerald-green wings make this dove distinctive. In female, white supercilium is less clear cut than in male.

Plain dove with black hindcollar. Can be distinguished from more common Spotted Dove (see p. 150), which is patterned on the wings and has a chequerboard pattern on hindcollar.

ORANGE-BREASTED GREEN-PIGEON
Treron bicinctus leggei

Size: 29cm

Habitat: Forested areas in lowlands.

Distribution: Large flocks sometimes gather in dry zone scrub forests.

Voice: Undulating whistle with chirruping notes interspersed. Has a musical quality to it and the notes are complex. Variations are uttered in different pitches. Similarly to Ceylon Green-pigeon, sings with theme of a whistled *ooo we ooo we oooo*, but vocalization is sharper and has quivering notes at the end of the whistle reminiscent of a motorbike that has been kick-started.

Status: Resident.

Male

Female

Both sexes have grey hindnecks (greenish in Ceylon Green-pigeon, see right). Undertail-coverts of latter are heavily marked in green.

Male

Male

CEYLON GREEN-PIGEON
Treron pompadora **e**

Size: 28cm

Habitat: Similar to Orange-breasted, but less common.

Distribution: Found throughout Sri Lanka.

Voice: Call a whistled *ooo we ooo we oooo* repeated with variations.

Status: Endemic.

Male has conspicuous purple-maroon mantle that female lacks. Female similar to Orange-breasted Green-pigeon (see left). Most easily told apart by having a greenish, rather than grey nape. Undertail-coverts also help to distinguish females of the two species; yellow in Ceylon Green-pigeon and cinnamon-red on inner webs of Orange-breasted.

Yellow-footed Green-pigeon
Treron phoenicopterus

Size: 33cm

Habitat: Tall forests in dry lowlands.

Distribution: Found in Bibile-Nilgala area.

Voice: Complex, undulating sequence of whistles.

Status: Race *philipsi* is a Scarce Resident. Critically Endangered on IUCN Red List. Race *chlorigaster* is a Highly Scarce Migrant.

Completely grey head and yellow feet distinguish it from similar female Orange-breasted Green-pigeon (see p. 152). Latter has a grey hindneck, but no grey on crown. Sexes similar, but female duller.

Green Imperial-pigeon
Ducula aenea pusilla

Size: 43cm

Habitat: Areas with tall forest. Favours canopy of tall trees.

Distribution: Found throughout Sri Lanka, except in far North. Visits Talangama Wetland near Colombo to feed on fruiting trees. Diet is mainly fruits, which are swallowed whole and hard seed is disgorged. Strong fliers that may cover large distances in a day.

Voice: Deep, throaty *room room room* booming from the canopy, sometimes escalating into an *oop oop oop* call.

Status: Resident.

Largest pigeon species resident in Sri Lanka. Has green upperparts.

PSITTACIFORMES
This order has a single family with more than 350 species, mainly in the tropics from Africa to Australasia.

Parrots Family *Psittacidae*
Parrots are deep-billed birds that use the bill as a clasping tool when clambering around. They utter harsh calls. Parrots and parakeets form large flocks. Hanging-parrots or lorikeets usually occur in pairs. Just one family is found worldwide, which is spilt into seven subfamilies.

CEYLON HANGING-PARROT
Loriculus beryllinus **e**

Size: 14cm

Habitat: Frequents tall forests. Fond of imbibing the sap of *Kithul* trees *Caryota urens*.

Distribution: Common in wet zone. Occurs locally in riverine forests in dry zone and certain dry zone areas such as Gal Oya.

Voice: Three-syllable, sharp, high-pitched call.

Status: Common Endemic.

Small green bird about the size of a House Sparrow (see p. 260). Adults have red crown and rump. Male more brightly coloured than female, which has only a trace of a blue throat. Juveniles have green head without red crown. In courtship, red rump feathers are raised.

ALEXANDRINE PARAKEET
Psittacula eupatria eupatria

Male

Size: 53cm

Habitat: Widespread in lowlands and hills. Needs tall trees for nest sites.

Distribution: Widespread except in highlands.

Voice: *Kraa* note often uttered in flight is deep and different from Rose-ringed Parakeet's, which sounds noisy.

Status: Resident.

Heavy red bill and red shoulder-patch distinguish this from Rose-ringed Parakeet (see p. 155). Large size and proportionately longer tail than in the other parakeets. Both sexes have large red shoulder-patch.

ROSE-RINGED PARAKEET
Psittacula krameri manillensis

Size: 42cm

Habitat: Adaptable bird that thrives in disturbed habitats and cultivated areas. Found in cities, where it excavates nest holes in tall, old trees. Readily visits bird tables for food.

Distribution: Widespread, especially in lowlands, where it is considered a pest by paddy farmers.

Voice: Screeching flight call.

Status: Common Resident.

Male has black collar edged with pink. Female has no collar.

Male

Male left, female right

Male

PLUM-HEADED PARAKEET
Psittacula cyanocepahala cyanocephala

Size: 36cm

Habitat: Usually found near thinly wooded forests adjoining open areas. Feeds on a variety of food from fruit, nectar and flower buds, to grain.

Distribution: Throughout Sri Lanka, but mainly in hills.

Voice: Sharp *eenk eenk* call often gives away its presence.

Status: Uncommon Resident.

Small size and plum head of male distinctive. Female has all-grey head. Both sexes have small red shoulder-patch.

LAYARD'S PARAKEET
Psittacula calthropae ⓔ

Size: 31cm

Habitat: Mainly in forested areas.

Distribution: In wet zone, up to mid-hills. Flocks also found in Nilgala area.

Voice: Raucous calls of a flock are distinctive. Two types of call are used regularly: a *we wik wur we wik wur* and a shrill, insistent *i-i-i-i-i-l*.

Status: Uncommon Endemic.

Male

Female has dark bill. Male has red bill tipped with yellow. Also called the Emerald-collared Parakeet, though 'emerald collar' is not always clearly seen in the field.

Female

CUCULIFORMES
Three families are placed in this order, of which the Hoatzin and turacos are found in South America and Sub-Saharan Africa respectively. The remaining family, the Cuculidae, is grouped into six subfamilies, of which three are found in Sri Lanka. These are the Old World cuckoos, coucals and malkohas. Birds in this order are zygodactylous, with two toes facing forwards and two backwards.

Cuckoos Family *Cuculidae*
This is a diverse family that includes the Old World cuckoos, as well as the malkohas and coucals, whose external appearances and behaviour are quite varied. The Old World cuckoos are the familiar cuckoos, and all of these species lay their eggs in the nests of unwitting hosts.

GREEN-BILLED COUCAL
Centropus chlororhynchos ⓔ

Size: 43cm

Habitat: Restricted to good-quality rainforests of large extent. Its survival in the relatively tiny Bodhinagala Rainforest is an anomaly. It may be lost from this reserve within the next few decades as urbanization continues.

Distribution: Restricted to a few lowland rainforests such as Sinharaja, Morapitiya and Kithulgala. A few also exist in small pockets such as Bodhinagala.

Voice: Call is *whoop-whoop whoop-whoop whoop-whoop*, or could even be transcribed as a *whoo whoo*. It is 'softer' sounding than call of common 'Southern' Coucal.

Status: Highly Scarce Endemic. Probably one of Sri Lanka's most endangered birds.

Looks very similar to 'Southern' Coucal (see p. 157). Green (not black) bill is diagnostic. Wings are duller and not bright chestnut as in 'Southern' Coucal. Black on body is glossier.

SOUTHERN COUCAL
Centropus sinensis parroti

Size: 48cm

Habitat: Found in a variety of habitats from forests to disturbed habitats in urban gardens. Often feeds on the ground. In urban habitats feeds on snails, and the presence of the invasive Giant African Snail *Lissachatina fulica* may be a key factor in it thriving in urban gardens.

Distribution: Widespread.

Voice: Broad range of vocalizations. Has a rapid and forceful *oop oop oop*. Some scolding calls. A throaty *kok* is uttered at times.

Status: Common Resident.

Black bird with stern, red eyes and chestnut wings. The old name of Crow-Pheasant captures its appearance and habits well. Previously known as Greater Coucal, a name now reserved for a species found in Northern India.

SIRKEER MALKOHA
Taccocua leschenaultii leschenaultii

Size: 43cm

Habitat: Lowland scrub jungle adjoining grassland.

Distribution: Dry lowlands in South-east of Sri Lanka. Seems to have very specific habitat requirements. In Yala National Park, for instance, Sirkeer Malkohas can be seen in very specific locations, unlike Blue-faced Malkohas (see p. 158), which can show up almost anywhere. They are often seen foraging on the ground.

Voice: Metallic *pip* that sounds like two pebbles being struck together.

Status: Uncommon Resident.

Plain brown, long-tailed malkoha with prominent red bill.

RED-FACED MALKOHA
Phaenicophaeus pyrrhocephalus Ⓔ

Size: 46cm

Habitat: Bird of tall rainforests. In the 18th century this bird was recorded by Captain Legge in Kotte on the outskirts of Colombo, which shows how rapid deforestation was in the 19th century.

Distribution: Mainly seen in a few remaining tall forests in lowland rainforests. Sinharaja, Morapitiya and Kithulgala are reliable for sightings. Scarce in dry zone. Mainly insectivorous, but does opportunistically feed on ripe berries.

Voice: Generally silent, but occasionally utters a guttural rattling call; almost a croak.

Status: Scarce Endemic. Vulnerable on IUCN Red List.

Male has brown irides, while female's irides are pale.

Male

Female

BLUE-FACED MALKOHA
Phaenicophaeus viridirostris

Size: 39cm

Habitat: Generally a shy bird of lowland scrub jungle.

Distribution: Dry lowlands. Feeds on a variety of plant and animal matter, including the '*Maliththan*' *Salvadora persica* fruit. Most of its diet probably consists of insects.

Voice: Like most malkoha species does not vocalize strongly, at times uttering a soft, guttural call in flight.

Status: Uncommon Resident.

In flight, white edges on long tail show up.

CHESTNUT-WINGED CUCKOO
Clamator coromandus

Size: 42cm

Habitat: Can occupy a range of wooded habitats from scrub forest and village gardens, to tropical rainforest. Most likely to be seen in good-quality forest.

Distribution: Spreads throughout Sri Lanka on arrival.

Voice: Double-noted *peep-peep* that is repeated. Reminiscent of a whistle being given two short, sharp blows. Usually silent in Sri Lanka.

Status: Uncommon Migrant.

JACOBIN CUCKOO
Clamator jacobinus jacobinus

Size: 31cm

Habitat: Forested areas in lowlands. Most likely to be seen in scrub jungle in dry lowlands. However, where it occurs in wet zone, often found in wetlands close to mangrove thickets.

Distribution: Widespread from lowlands to mid-hills.

Voice: Repetitive, metallic piping call.

Status: Uncommon Resident.

Chestnut wings, white underparts, and black on face, crest, back and tail. Large, distinctive cuckoo. Despite the bold colouring, can be easily overlooked as it does not draw too much attention to itself.

Adults are unmistakable in black and white. Juveniles look like browner version of adults. In flight they show white wing-patches.

ASIAN KOEL
Eudynamys scolopaceus scolopaceus

Size: 43cm

Habitat: Favours lightly wooded areas with fruiting trees. Frequents gardens in cities. Very partial to fruits of palms and ripening papayas.

Distribution: Found throughout Sri Lanka, except highlands.

Voice: Series of *ko wuu* calls. About 7–8 notes are usually uttered, very loudly, each rising in pitch and sounding more insistent. Notes can change as they become somewhat tremulous. May also rapidly utter three *ko wuu kowuu ko wuu* notes of even pitch. The Sinhala name *Koha* is somewhat onomatopoeic, and derives from these *ko wuu* notes. Can also produce a cackle of a few *kik kik* notes. I once saw a male call for at least half an hour on and off, while another male was perched just a foot away.

Status: Common Resident.

Male

Female

Male looks similar to a crow in glossy black plumage. However, ivory-coloured bill, red eyes and slimmer shape distinguish it from crows. Female is brown, heavily barred and spotted with white.

ASIAN EMERALD CUCKOO
Chrysococcyx maculates

Female

Size: 18cm

Habitat: Found in mix of wooded habitats from primary and secondary forests, to village gardens. In Sri Lanka recorded in monsoon forests in dry lowlands around Sigiriya.

Distribution: May potentially turn up anywhere, including in mountains, as in its breeding range it has a wide elevation range.

Voice: Repeated, double-noted *jit-huyeet jit-huyeet*. Also trilling call.

Status: Vagrant.

Male metallic green on head, body and wing-coverts, with barred black-and-white belly. Female has duller barring on underparts, which extends from throat to vent. Barring extends to cheeks and ear-coverts, with crown and nape being orange-brown. Juvenile heavily barred, similar to female.

BANDED BAY CUCKOO
Cacomantis sonneratii waiti

Size: 24cm

Habitat: Favours tall forests in both wet and dry zones.

Distribution: Lowlands to mid-hills where sizeable forest patches remain.

Voice: Double-noted *pee-phew pee-phew*, repeated, sometimes in a sequence of two, three or four of these double-noted calls. Clear, clean and musical. Calls may sometimes be preceded by series of rising single notes.

Status: Uncommon Resident.

Upperparts are brown, barred with darker brown. Underparts white barred with thin brown bars. Can look similar to female and hepatic phase of Small Cuckoo (see p. 163), but Banded Bay has pale face with thick brown mask running from eye to ear-coverts. Face and head of Lesser are reddish-brown with barring, and lack facial mask. Juvenile lacks complete dark crown and has pale edges to wing-coverts. Race *waiti* in Sri Lanka is confined to the island.

Male   Female

GREY-BELLIED CUCKOO
Cacomantis passerinus

Size: 23cm

Habitat: Most likely to be seen in areas of scrub forest.

Distribution: Widespread winter visitor, most likely to be seen in dry zone scrub forests, but also found in wet zone and lower hills.

Voice: Repeated whistled *tcheow*.

Status: Migrant.

Adult male is grey. Female also has brick-red (or hepatic) form. Juvenile also brick-red, but has diamond markings on tail that distinguish it from hepatic phase female. Juvenile also has less rufous on throat and breast, and is more heavily barred. Rufous morph of Small Cuckoo (see p. 163) has unmarked rump and nape. In grey-phase birds white edge on shoulder may be a useful field identification characteristic at times. Tail can look pointed or squarish. Previously this species was also known as the Plaintive Cuckoo, a name now reserved for C. merulinus found in Southeast Asia that migrates to India, but is not recorded in Sri Lanka. However, the change of name causes confusion as the older literature uses the name Plaintive Cuckoo.

FORK-TAILED DRONGO-CUCKOO
Surniculus [lugubris] dicruroides

Size: 27cm

Habitat: Found in secondary forests in wet and dry zones. Occurs where forests border village gardens.

Distribution: Found throughout Sri Lanka from lowlands to highlands. Absent in arid scrub forests in coastal areas.

Voice: Typically a series of six notes of rising intonation that are repeated. Each note is brief and spacing between notes is so short that it is hard to make out that there are as many as six notes in each sequence. Distinctive call. May imitate *chowck chowck* calls of Ceylon Crested Drongo (see p. 272), which raises interesting questions as to whether this may be related to it laying its eggs in the nests of the drongos and cuckolding them.

Status: Uncommon Resident.

This cuckoo can be mistaken for a drongo. However, weak bill and white barring on vent and underside of tail feathers help to distinguish it. Also, both races of White-bellied Drongo (see p. 271) have a white vent. Drongo-cuckoo has dark vent with white barring.

Male

COMMON HAWK-CUCKOO
Hierococcyx varius

Size: 34cm

Habitat: Prefers wooded areas with tall, mature trees. Occasionally strays into plantations adjoining forests.

Distribution: From mid-hills to highlands. Migrant race may be seen in lowlands before it continues to higher elevations.

Voice: Tremulous, double-noted call likened to 'Brain Fever'. Can increase pace of call and sound slightly hysterical.

Status: Race *ciceliae* is an Uncommon Resident, and race *varius* is a Highly Scarce Migrant.

Grey upperparts and distinctively barred tail. Tail barring is crenulated with adjoining dark and thin pale bars, with generous spacing between them. Orange breast-band extends to throat. In some birds throat is streaked. Yellow eye-ring. Confusion with other cuckoos in Sri Lanka is unlikely.

SMALL CUCKOO *Cuculus poliocephalus*

Size: 26cm

Habitat: Favours forest patches, though can tolerate more open areas with village gardens.

Distribution: On arrival, spreads across lowlands up to higher hills.

Voice: Repeated sequence of varying notes, *whip-kik-kik-ke-keuw*.

Status: Scarce Migrant.

Female   *Male*

INDIAN CUCKOO *Cuculus micropterus*

Size: 33cm

Habitat: Well-wooded areas. In dry lowlands occupies more patchy habitats as long as they contain groves of mature trees.

Distribution: Resident race found largely in dry lowlands on an arc from North to South-east. Migrant race spreads across wet zone as well as up to mid-hills.

Voice: Clean and clear *ku-ku-cuckoo* that is sometimes repeated for several minutes without pause.

Status: Uncommon Resident and Migrant.

Small Cuckoo's tail is blackish and is the best identification feature in an adult male. Indian Cuckoo (see right) has broad black subterminal band on grey tail. Indian female has more buff on throat and upper breast than grey morph female Small. Juvenile has pale edges to wing-coverts and pale spots on brownish wings. Grey morph juvenile has blackish head. Rufous morph of female Small may be confused with Grey-bellied (see p. 161) juvenile and red-phase female. Rump and back of head of Grey-bellied juvenile is unmarked.

Indian Cuckoo and Small Cuckoo (see left) are both smaller than Common Cuckoo (see p. 164). Indian has broad black subterminal band on grey tail. Small Cuckoo's tail is blackish. Indian female has more buff on throat and upper breast than grey morph female Small. Juvenile has pale edges to wing-coverts. Female has no hepatic phase.

COMMON CUCKOO Cuculus canorus

Size: 33cm

Habitat: Open country with wooded groves. In breeding range can be found on farmland, as long as it contains little wooded copses.

Distribution: Spreads from lowlands to mid-hills.

Voice: When wintering in Sri Lanka, the famous cuckoo call is not heard. Easily overlooked as it is silent.

Status: Scarce Migrant.

Fine black barring on underparts. Male has dark grey tail contrasting with rump and upperparts. Tail contrast less pronounced compared with Small Cuckoo's (see p. 163). Female has a grey phase and a hepatic phase.

STRIGIFORMES
Two families are included in this order of birds. They are familiar but rarely seen in the wild by many people on account of their mainly nocturnal habits.

Barn-owls and bay owls Family *Tytonidae*
These owls are characterized by a heart-shaped face and are generally medium in the size range of owls.

COMMON BARN-OWL Tyto alba

Size: 36cm

Habitat: Most encounters people have are of barn-owls roosting in abandoned buildings or disused lofts. I have even seen this owl coming to the lawn of the Galle Face Hotel in Colombo, and seen photographs taken in Ladies College in the heart of Colombo. Hunts in open areas for rodents, but has adapted to urban settings.

Distribution: Found throughout lowlands.

Voice: Throaty hiss. Not very vocal.

Status: Scarce Resident.

Ghostly appearance with white underparts and pale orange-yellow upperparts that are intricately patterned. Heart-shaped white facial disc makes this bird easy to identify. Short-eared Owl (see p. 170) also has heart-shaped facial disc, but it is not white and confusion between the two species is unlikely.

CEYLON BAY OWL *Phodilus assimilis*

Size: 29cm

Habitat: Known only from good-quality forests.

Distribution: Found in wet zone up to highest mountains, and also in intermediate zone.

Voice: Three-syllable whistle that sounds a little like *oh-kee-yow*. Slightly tremulous.

Status: Highly Scarce Resident. One of the least seen bird species. It was not seen in the wild in Sri Lanka until 2001.

Barn-owl-like triangular facial disc with dark crown. Flight feathers are barred and mantle feathers have lines of black-and-white dots. Upperparts darker with a hint of chestnut. Stout build.

> **Owls** Family *Strigidae*
> This family comprises owls that are not barn-owls or bay owls. They are varied in behaviour and not all species are nocturnal.

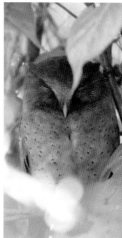

SERENDIB SCOPS-OWL
Otus thilohoffmanni **e**

Size: 17cm

Habitat: Disturbed forest edges, often near the ground. Appears to be insectivorous.

Distribution: Restricted to a few lowland rainforests in Sri Lanka, such as Sinharaja, Morapitiya and Kithulgala.

Voice: Call is a soft *phew*.

Status: Scarce Endemic. First seen in 2001. Described to science only in 2004 by Deepal Warakagoda (who discovered it) and Pamela Rasmussen. The first endemic bird to be described from Sri Lanka after a lapse of 32 years. It is amazing that an endemic bird in Sri Lanka eluded being known for so long. Its scientific name honours Thilo Hoffmann, who is credited with leading the campaign to halt the destruction of the Sinharaja Rainforest.

Overall reddish hue and soft, single-note call help make confusion between species unlikely. Male has orangeish irides; yellowish in female. Pairs are believed to maintain a territory year round.

ORIENTAL SCOPS-OWL *Otus sunia*

Size: 19cm

Habitat: Found in thick forest in hills and scrub jungle in dry lowlands. Roosts on tops of trees as well as in bushes.

Distribution: Mainly in dry lowlands and can ascend to higher hills. Absent in wet lowlands.

Voice: Repeated *wuk-krrrrr*, or sometimes just a repeated *krrrr*. Last note reverberates.

Status: Scarce Resident.

INDIAN SCOPS-OWL
Otus bakkamoena bakkamoena

Size: 24cm

Habitat: Village and town gardens, and secondary forest. Has adapted to urban habitats.

Distribution: Throughout Sri Lanka from lowlands to highlands. The most common and most widespread of the owls.

Voice: Very vocal during breeding season. Birds utter a *whuk whuk* call. Sometimes a series of notes is uttered rapidly.

Status: Resident.

Occurs in grey and rufous phases. Best distinguished from Indian Scops-owl (see right) by yellow irides (brown in Indian). Also lacks pale hindcollar and has well-marked underparts. Smaller than Indian. Scapular stripe whitish (buff in Indian).

Brown irides in Indian Scops-owl help to distinguish it from rarer Oriental Scops-owl (see left). Two colour forms, one greyish and the other rufous. Indian is marked less on underparts than Oriental. I have noticed that some of the grey-form birds are less marked than the rufous forms. Ear-tufts of Indian are well formed and often erected by the bird.

SPOT-BELLIED EAGLE-OWL
Bubo nipalensis blighi

Size: 63cm

Habitat: Frequents tracts of large forests, but most sightings by birders are from where the owls have roosted close to human habitations adjoining forests.

Distribution: Found throughout Sri Lanka up to highlands. Most records are from large tracts of lowland dryzone and wet zone jungles.

Voice: Whistle that sounds like a nasal and slightly explosive *kee yow*.

Status: Uncommon Resident.

Large owl with prominent 'horns' on head. Also referred to as Forest Eagle-owl. Underparts have downwards-pointing arrowheads. Flight feathers are barred and wing-coverts are patterned in the middle, lending an overall impression of barring and mottling. Juvenile paler than adults.

Juvenile

BROWN FISH-OWL
Ketupa zeylonensis zeylonensis

Size: 61cm

Habitat: Widespread up to mid-hills. Tolerant of human habitation.

Distribution: Widespread from lowlands to highlands. Grounds of Cinnamon Lodge and Chaaya Village at Habarana are reliable sites. The owls are often seen roosting on large *Kumbuk* trees adjoining freshwater lakes such as at Habarana and Wilpattu.

Voice: Double-noted, deep *uhm-ooom*.

Status: Resident.

Large size, prominent yellow eyes and distinct thin streaks on underparts. Upperparts heavily marked and wings are barred. Partly diurnal and the owl most seen by visitors to national parks.

BROWN WOOD-OWL
Strix leptogrammica

Size: 50cm

Habitat: Favours forests with mature trees. Not shy of roosting close to human habitation. May be found on estates that are a patchwork of old forest and degraded cultivated land.

Distribution: Found throughout Sri Lanka where suitable wooded areas remain.

Voice: Deep, far-sounding *hu-hu-hoo*.

Status: Uncommon Resident.

Large size, and dark brown eyes bordered with dark brown set in rich buff facial discs. Dense, thin barring on underparts. Confusion unlikely with any other owl.

JUNGLE OWLET
Glaucidium radiatum radiatum

Size: 20cm

Habitat: Mainly lowland scrub jungles.

Distribution: Dry zone. Seems to be abundant in Moneragala-Nilgala area.

Voice: Some call notes similar to Chestnut-backed Owlet's. Series of almost barking-like *ooh ooh* notes that are slightly tremulous. Alternates with more insistent barking *oo oo* note.

Status: Resident.

Small owl that is barred black and white. Face and underparts similar to Chestnut-backed Owlet's (see p. 169), but confusion is unlikely when upperparts are seen. Active by day.

CHESTNUT-BACKED OWLET
Glaucidium castanonotum ⓔ

Size: 19cm

Habitat: Well-wooded areas, though it is not shy of roosting near human habitation where it adjoins good-quality forest.

Distribution: Confined to wet zone. Probably more common than it is believed to be. Interestingly, Jungle and Chestnut-backed Owlets appear to carve out different areas in Sri Lanka.

Voice: Higher pitched than Jungle Owlet's, and more like a throaty, tremulous *kruk kruk*. Also a repeated, single-noted *wheeuw*, which it alternates in pitch from one note to another other and repeats.

Status: Endemic. Vulnerable on IUCN Red List.

Face and underparts similar to those of Jungle Owlet (see p. 168), but chestnut upperparts easily separate the two species.

BROWN HAWK-OWL
Ninox scutulata hirsuta

Size: 32cm

Habitat: Has adapted to urban environments and uses TV antennae as a substitute for tall perches. In cities such as Colombo, I have often traced a calling bird to a TV antenna that is used as a boundary perch.

Distribution: Widespread throughout Sri Lanka.

Voice: Distinctive *ku wook ku wook* call that is repeated. I have heard it uttering the same call softly, making it seem much further away than it really is. Rarely, I have also heard it uttering a yelping call.

Status: Resident.

About the size of a crow, with largely plain upperparts. Underparts white with lines of heavy rufous spots, leaving a few areas of white on breast. All white around vent.

SHORT-EARED OWL *Asio flammeus*

Size: 38cm

Habitat: Favours open areas including marshes.

Distribution: May turn up anywhere in Sri Lanka. I have seen it in the marshes in Colombo.

Voice: Call sounds like a snarl made by a small animal, *rrrr-row*.

Status: Scarce Migrant.

CAPRIMULGIFORMES
There are five families in this order, with some being restricted to certain geographical areas. Two families are found in Sri Lanka.

Frogmouths Family *Podargidae*
Frogmouths are nocturnal, wide-mouthed birds with a large bill and conspicuous rictal bristles around it. They use cryptic colouring for camouflage, which makes them look a little like broken twigs. The family has two genera, one in Asia, and the other in Australia and New Guinea.

CEYLON FROGMOUTH
Batrachostomus moniliger

Size: 23cm

Habitat: Well-forested areas in lowlands.

Distribution: Widespread in lowlands where remnant patches of forest are found. Can occupy degraded habitats, but only if they adjoin good-quality forests. A subcontinental endemic found in India in the Western Ghats.

Voice: Harsh, explosive *whaa*, descending in tone.

Status: Resident.

Yellow irides, and dark mask-patch around eyes. Barring on upperparts is not tight and linear, and creates an impression of rich patterning. Thick, dark streaking on underparts. Underwing pale except for dark tips on outer primaries and dark marking on wrist.

Female is always rufous. Males occur in grey and brown colour phases (morphs). Male identified by pale, 'lichen-like' patches on lower scapulars and tertials. A reliable way to distinguish the sexes was not known until Deepal Warakagoda published a paper in the *Ceylon Bird Club Notes*.

> **Nightjars** Family *Caprimulgidae*
> Nightjars are nocturnal birds, similar in shape and all cryptically camouflaged to avoid detection when roosting on the ground. A few species perch on trees between periods of activity or when calling. Some species alight on the ground between bouts of hunting or when calling.

GREAT EARED-NIGHTJAR
Eurostopodus macrotis

Size: 39cm

Habitat: Found in wet zone forests in lowlands and lower hills in India.

Distribution: A record from Sri Lanka was also from the lowland wet zone.

Voice: Loud, clear, repeated whistle, *pee-wheew*.

Status: Vagrant.

Noticeably larger sized than the other nightjars, which is obvious when a size comparison is possible. Long tail. Absence of white in tail and primaries also separates it from the other three nightjars. White throat-patch and pale band on lower belly. Male has black spots on cinnamon-tinged grey cap, and no longitudinal streaks as in the other three species. Dark face, throat and breast.

Telling Apart Jerdon's, Indian Little & Indian Jungle Nightjars

	Jerdon's	Indian Little	Indian Jungle
Size	Babbler-sized	Bulbul-sized	Babbler-sized
White throat- band	Unbroken	Broken	Broken in male, absent in female
Pale moustachial stripe	Yes	Yes	Yes
Tail	Long	Short	Medium length
White tips to tail feathers	Large	Large, almost half of tail	Small
Crown	Streaked	Very strongly streaked	Streaked
Overall colour	Browner, warmer	Greyer, darker	Grey
Call	Tremulous	Like marble being dropped	Double-noted call

INDIAN JUNGLE NIGHTJAR
Caprimulgus indicus

Size: 29cm

Habitat: Found in scrub jungle and grassland.

Distribution: G. M. Henry reported that this species was found mainly above elevations of 1,000m, but most recent records are from dry zone in northern and eastern areas of Sri Lanka.

Voice: Rapidly repeated, resonant, double-note *chuck-koo chuck-koo.*

Status: Resident.

Also referred to as the Grey Nightjar or Highland Nightjar. Grey Nightjar is now used for *C. jotaka* (not recorded in Sri Lanka). As in Jerdon's and Indian Little Nightjars (see right and p. 173), it also has a white patch on primaries and at terminal sides of tail, but white patch on primaries is smaller. Male has two white spots on throat and pale moustachial stripe. In female white throat-patch is replaced with warm buff. First-year male similar to female with pale, almost unmarked tertials.

JERDON'S NIGHTJAR
Caprimulgus atripennis aequabilis

Size: 26cm

Habitat: Resident of dry lowlands, especially in areas of open scrub, but Jerdon's often calls perched from a vantage point on a tree or bush.

Distribution: Mainly in dry zone from lowlands to mid-hills, with a few populations in wet zone. A few birds have been recorded in wet zone during migrant season, getting lost and going into houses (I photographed one such bird in Colombo). This may indicate that there might be a 'local movement', or possibly an influx of migratory birds.

Voice: Quivering, double-noted *ku-krr,* with a liquid quality.

Status: Resident.

Male has a single white patch (also referred to as a white gorget) on throat. Indian Little Nightjar (see p. 173) has white gorget broken in the middle, creating two white throat-patches. Both Jerdon's and Indian Little Nightjars have mantle feathers with pale edges and greater coverts also with pale edges, giving rise to wing-bars. However, Jerdon's Nightjar is a little more contrastingly plumaged than other nightjar species. It has a longer tail than Indian Little. Jerdon's Nightjar is appreciably larger than Indian Little, and is the size of a Yellow-billed Babbler (see p. 237). Size and call may provide the best ways to distinguish the species.

INDIAN LITTLE NIGHTJAR
Caprimulgus asiaticus eidos

Size: 24cm

Habitat: Open areas in dry lowlands.

Distribution: Mainly in dry zone from lowlands to mid-hills. A few populations in wet zone.

Voice: *Chuk chuk chukoor* with the last note being tremulous. Likened to the sound of a marble being dropped. Another descriptive name for this bird is Marble Dropper.

Status: Resident.

Similar to Jerdon's Nightjar (see details under Jerdon's, p. 172). Indian Little is smaller and approximately the size of a Red-vented Bulbul (see p. 212). Indian Little has a strongly streaked crown (Jerdon's is also streaked), and a buff hindneck-patch. White gorget (or throat-bar) is broken in the middle (Jerdon's is unbroken). Central tail feathers are unmarked (marked in Jerdon's). Differences in calls provide the best way to distinguish the two species unless a size comparison can be made.

APODIFORMES
There are three families in this order, the swifts, treeswifts and hummingbirds, with the latter being confined to the New World.

Swifts Family *Apodidae*
Swifts are insectivorous, narrow-winged birds that spend almost their entire life on the wing, including when sleeping. The fastest flying species have long, swept-back wings.

INDIAN SWIFTLET
Aerodramus unicolor

Size: 12cm

Habitat: Generally found not far from good-quality forests, though it can sometimes nest in urban environments.

Distribution: Found in wet zone, ascending to highlands.

Voice: High-pitched, chittering calls. Sounds very similar to Little Swift.

Status: Resident.

Plain colour distinguishes this species from Little Swift (see p. 176), which has a white rump. Its shorter tail distinguishes it from Asian Palm-swift (see p. 174). Indian Swiftlet has a distinctive flight, gliding with its wings stiffly held parallel to the ground. Glides last only a few seconds and are interspersed with a rapid flutter of the wings.

Brown-throated Needletail
Hirundapus giganteus

Size: 23cm

Habitat: Found in small flocks over densely forested areas in wet zone. This may be a reflection of its preferred insect prey, or it may be limited by the need for undisturbed nest sites.

Distribution: Restricted to wet zone forests in the South-west. Most likely to be seen where the forests have a hilly aspect.

Voice: High-pitched *chirrup*, or sequence of sharp squeaks.

Status: Scarce Resident.

Overall impression is of a dark, broad-winged swift flying very fast. White flanks and undertail-coverts separate it from Alpine Swift (see p. 175), but they do not show well in some conditions. Upperparts uniformly dark.

Asian Palm-swift
Cypsiurus balasiensis balasiensis

Size: 13cm

Habitat: Most common in lowlands up to mid-hills. Flocks may move around, but are 'centred' around a grove of palm trees most of the time.

Distribution: Widespread throughout Sri Lanka.

Voice: Brief chittering call, high in pitch, uttered frequently as a flock swoops around.

Status: Resident.

May be confused with Indian Swiftlet (see p. 173), previously known as Edible-nest Swift, and distinguished from it by the comparatively long wings and long tail. Tail is sporadically spread open, showing a fork. Overall impression is of a greyish-brown bird.

ALPINE SWIFT *Tachymarptis melba*

Size: 22cm

Habitat: Alpine Swifts are great travellers and can turn up anywhere. I have even seen them in Yala National Park in the South-east, but they are more typical of mountainous areas.

Distribution: Recorded from all over Sri Lanka, but stronghold is in higher hills and highlands. Flocks in lowlands are likely to be feeding parties that have arrived temporarily from hills.

Voice: Rapid, machine gun-like sequence of short, sharp notes that rise and fall in pitch.

Status: Uncommon Resident.

PACIFIC (FORK-TAILED) SWIFT
Apus pacificus

Size: 18cm

Habitat: In its breeding range in North Asia, found in mountains.

Distribution: May turn up anywhere in Sri Lanka, but most likely to occur in hilly habitats favoured by Alpine Swifts (see left).

Voice: High-pitched, thin *kree-kree* call that sounds squeaky.

Status: Vagrant.

Largest of the swifts in Sri Lanka. Long wings are bowed back. Brown band separates white throat from white belly. No white on undertail-coverts as on Brown-throated Needletail (see p. 174), which also has a dark throat. Fast flight, often in flocks.

Easily overlooked for more common Little Swift (see p. 176). However, it is slimmer bodied and longer winged, and has a diagnostic deeply forked tail, lacking in Little. Underparts are scaly. In Little, underparts are uniformly dark.

LITTLE SWIFT *Apus affinis singalensis*

Size: 15cm

Habitat: Usually near rocky outcrops, where it nests in caves or overhangs. May also nest in old buildings.

Distribution: Occurs in large flocks throughout Sri Lanka up to mid-hills.

Voice: High-pitched, loud, screeching call made on the wing as it careers around.

Status: Resident.

White rump against black plumage is distinctive (previously known as White-rumped Swift). White throat may not show in flight. Underparts often look plain black, and general impression is of an overall black swift until white rump is seen (see also description for Pacific Swift, p. 175).

Treeswifts Family *Hemiprocnidae*
These narrow, long-winged, insectivorous birds spend a lot of time in the air. They perch on trees and do not use wires like swallows do. The four species of treeswift are confined to the region between South Asia and the Solomon Islands.

CRESTED TREESWIFT
Hemiprocne coronata

Male (left) Female (right)

Female

Size: 23cm

Habitat: Widespread in lowlands and lower hills. Most likely to be encountered close to forest patches.

Distribution: Widespread except in highlands. Absent in urban areas.

Voice: Nasal, double-note *kee-yeew*, usually uttered on the wing.

Status: Resident.

Overall impression is of a large, grey, swift-like bird with long wings and long tail. Not as highly manoeuvrable in the air as swifts, and perches frequently on branches. Male's ear-coverts are red; female's are dark grey, almost black. Both sexes have tall crest at base of bill that is raised when perched.

MALABAR TROGON
Harpactes fasciatus fasciatus

Size: 31cm

Habitat: Good-quality forests in lowlands and hills. It still holds out in secondary forest fragments close to areas that have been cleared and degraded by farming.

Distribution: Found throughout Sri Lanka where forest cover is present. Numbers are highest in wet zone.

Voice: Soft, single-note, repeated *phew* that is slightly sibilant and whistle-like.

Status: Uncommon Resident.

TROGONIFORMES
This order contains only one family, the trogons, which are more numerous in the New World than the Old World.

Trogons Family *Trogonidae*
These are sexually dimorphic birds, with the males being strikingly coloured and the females drab. They usually occur in pairs.

Male has red underparts with white collar separating black head. Female has brown head contrasting with underparts that are brownish with a hint of orange. Both sexes have wing-coverts barred in black and white.

Male

Female

CORACIIFORMES
Of the nine families in this order, five are found in Sri Lanka. The families show a huge variety, including birds such as the diminutive Common Kingfisher and the large Malabar Pied Hornbill.

Kingfishers Family *Alcedinidae*
These are long-billed birds. Not all of them are predominantly fish eaters; some are forest kingfishers. They are omnivorous. The smaller species have high-pitched calls.

COMMON KINGFISHER
Alcedo atthis taprobana

Size: 17cm

Habitat: Widespread in lowlands and hills. Never far from water and predominantly feeds on small fish. Will also take other aquatic animals, including small crabs and amphibians.

Distribution: Widespread, from lowlands to highlands,

Voice: High-pitched, metallic, repeated *tee tee* call that varies a little in pitch. Call often announces its arrival as a tiny blue dart skimming over the water.

Status: Resident.

Male

Both mandibles are black in male. Female has orange on most of lower mandible. Orange ear-coverts are diagnostic and distinguish this species from Blue-eared Kingfisher (see right). Lighter, bright blue stripe running along back.

BLUE-EARED KINGFISHER
Alcedo meninting phillipsi

Size: 16cm

Habitat: Forested streams and rivers.

Distribution: Most records are from dry lowlands. Unusually, there are no photographically confirmed identifications, even given the explosion of digital wildlife photography in recent years.

Voice: Thin, high-pitched *cheee chee*. Less distinct than Common Kingfisher's call.

Status: Highly Scarce Resident. Highly Endangered on IUCN Red List. Some individuals may be migrants from Asia.

Male

Diagnostic feature that distinguishes this species from Common Kingfisher (see left) is blue ear-coverts, as opposed to orange ones in Common. Upperparts are deeper blue and underparts are deeper orange-chestnut. Common can at times hunch its head, so that orange ear-coverts do not show. This can lead to erroneous records. Sometimes seeing a Common Kingfisher close-up in the field or in good lighting conditions may make it look different (I have examined many small kingfishers and found them all to be Common). Many records of Blue-eared Kingfishers in Sri Lanka are erroneous. The image here is of one photographed in Borneo. I have seen Blue-eared in Peninsular Malaysia and Borneo, but not yet in Sri Lanka. Male has black bill and female has orange lower mandible base.

BLACK-BACKED DWARF KINGFISHER
Ceyx erithaca erithacus

Size: 13cm

Habitat: In forested streams in lowlands up to mid-hills.

Distribution: From lowlands to mid-hills.

Voice: Thin, whistled *hee hee*, similar to Common Kingfisher's.

Status: Uncommon Resident.

Distinctive small kingfisher with orange body and blackish-blue wings. Easily overlooked because of small size (sparrow sized) and discreet habits. Its presence is best noted from its call.

STORK-BILLED KINGFISHER
Pelargopsis capensis capensis

Size: 38cm

Habitat: Riverine habitats and freshwater lakes in lowlands and hills. In cities like Colombo, which have many aquatic habitats in the suburbs, may visit gardens and take invertebrates and small reptiles, but its preferred habitat is by water.

Distribution: Widespread up to mid-hills.

Voice: Rising *kee kee* changing into mournful wailing series of notes followed by laughing calls.

Status: Uncommon Resident.

Large kingfisher with large red, dagger-like bill. Yellowish-buff body, with pale chocolate crown, and blue wings, mantle and tail.

WHITE-THROATED KINGFISHER
Halcyon smyrnensis fusca

Size: 28cm

Habitat: Often seen on telegraph wires over paddy fields.

Distribution: Found throughout Sri Lanka from lowlands to highlands. Hunts invertebrates, which form the bulk of its diet. Will take fish, but is more of a forest kingfisher. Invertebrate diet brings it to gardens even in Colombo.

Voice: In flight, a harsh, repeated *kra kra*; whinnying call at rest. Nasal chink is also uttered from a perch.

Status: Common Resident.

White throat and breast contrast strongly with chocolate underparts and head. Red bill, and blue upperparts with chocolate wing-coverts. In flight shows white patches formed by bases to primaries.

BLACK-CAPPED KINGFISHER
Halcyon pileata

Size: 30cm

Habitat: Rivers, lagoons and mangrove habitats in lowlands. One bird took up residence by a bridge on the busy A2, near Kalutara; another visited Uda Walawe National Park for at least three consecutive years.

Distribution: Recorded mainly in lowlands.

Voice: Similar to White-throated Kingfisher, but calls are deeper.

Status: Scarce Migrant.

Black cap and broad white collar separating from purple upperparts and orange underparts. Red bill and black wing-coverts. Unlikely to be confused with White-throated Kingfisher (see left) in the same genus.

LESSER PIED KINGFISHER
Ceryle rudis leucomelanura

Size: 31cm

Habitat: Favours brackish water habitats in coastal areas, such as lagoons and estuaries. Also found in freshwater marshes and rivers. Abundant in mangrove habitats such as Muthurajawela marshes north of Colombo.

Distribution: Found throughout lowlands, but more common close to coastal strip. Birds occasionally visit or are resident in freshwater marshes and lakes in the interior.

Voice: Chirruping call similar to that of a tern.

Status: Uncommon Resident.

Black-and-white kingfisher with a habit of hovering over water. Female has a single breast-band, male has two. Black bill and crest.

Bee-eaters Family *Meropidae*
These insectivorous birds sally forth from high vantages after their insect prey. They have thin, long bills and long wings.

Air-space Niche Partitioning in Bee-eaters
Similar species can occupy the same physical environment if they have divided up the physical and biological habitat into niches that they occupy and use to feed in. Sometimes the niche partitioning is subtle, and may be related to factors such as water salinity or length of immersion in tidal waters. The three common species of bee-eater in Sri Lanka reveal an obvious physical separation in the air space they chose to hunt in, though there can be some overlap. Little Green Bee-eater perches low, sometimes on a clod of earth, a pile of dung or a low branch in open areas. Chestnut-headed Bee-eater perches in the mid-canopy in forests, and is unlikely to be seen in open spaces. The migrant Blue-tailed Bee-eater hunts high in the air and perches high up in the canopy. At the edge of thorn-scrub forest with tall trees in a park like Yala, it may be possible to see all three species occupying their preferred vertical zone. However, this is a generalization and Blue-tailed Bee-eaters are infrequently seen perched low in national parks, but never in city habitats.

Male

LITTLE GREEN BEE-EATER
Merops orientalis orientalis

Size: 21cm

Habitat: Scrub forest bordering open areas. Like all bee-eaters, takes insects on the wing.

Distribution: Dry lowlands.

Voice: Shrill *chirp*, sharper in tone than that of other bee-eaters.

Status: Resident.

Black bill forms continuous black line passing through eye to ear-coverts. This and middle tail-streamers are shared with Blue-tailed Bee-eater (see right), with which confusion is possible. Blue-tailed has a 'tricolour' pattern below black eye-to-ear line, with a thin blue line separated by yellow and buffy-chestnut areas. 'Blue tail', if seen, also distinguishes larger Blue-tailed Bee-eater. Adult Little Green Bee-eater has a greenish-blue throat demarcated at the bottom with a thin black line. Juveniles have green throats. Often perches low.

BLUE-TAILED BEE-EATER
Merops philippinus philippinus

Size: 31cm

Habitat: Chooses tall perches, be it trees or television antennae in cities. Prefers to hunt over open spaces such as paddy fields or scrub forest bordering grassland, the exception being cities, where it hunts at a good height over roof level, taking insects on the wing.

Distribution: Spreads all over Sri Lanka.

Voice: Call is a cheerful twittering, uttered at rest and in flight.

Status: Common Migrant.

Large size, and blue tail with elongated central feathers (see also details under Little Green Bee-eater, left).

EUROPEAN BEE-EATER *Merops apiaster*

Size: 27cm

Habitat: Forests bordering open areas.

Distribution: In the last few years, flocks have been recorded in Yala National Park.

Voice: Rapidly repeated, quivering *peek peek* notes.

Status: Scarce Migrant.

CHESTNUT-HEADED BEE-EATER
Merops leschenaulti leschenaulti

Size: 21cm

Habitat: Forest patches. Perches at mid-level of trees.

Distribution: Throughout lowlands to mid-hills. Low in number and not seen in big flocks like the other three bee-eaters occurring in Sri Lanka.

Voice: Tremulous *wheow wheeow* similar to Blue-tailed Bee-eater's.

Status: Uncommon Resident.

Blue underparts and extensive chestnut on wings distinguish this species from Chestnut-headed Bee-eater (see right). Crown and back of head are chestnut in both species. Chestnut-headed has a blue rump but lacks blue underparts. European also has extensive chestnut on shoulder area of wing. In Chestnut-headed, chestnut areas are confined to crown and mantle.

Blue rump in flight. Lacks elongated central tail feathers found in other bee-eaters recorded in Sri Lanka. Migrant European Bee-eater (see left) has blue underparts.

> **Rollers** Family *Coraciidae*
> Rollers are omnivorous, crow-sized, stockily built birds that are brightly coloured. They get the name roller from their tumbling courtship flights.

EUROPEAN ROLLER *Coracias garrulus*

Size: 31 cm

Habitat: Bird of open country interspersed with tall trees providing a high vantage.

Distribution: May turn up anywhere. First record of it was in Uda Walawe National Park.

Voice: Rolling, rattling calls and some tremulous *tchowk* notes. Some calls are reminiscent of more common Indian Roller's, but recognizably different.

Status: Vagrant.

Differs from resident Indian Roller (see right) by more uniform underparts lacking pinkish breast. Flight feathers are black. In Indian, there is a lilac band across rear edge of secondaries that continues across middle of primaries. Tail in European Roller is uniform, while in Indian a lilac-coloured T is formed. European has cinnamon-brown mantle and tertials.

INDIAN ROLLER
Coracias benghalensis indica

Size: <31 cm

Habitat: Open areas in lowlands. Feeds mainly on insects taken on the wing. Often hawks for insects that take flight when a forest is burnt.

Distribution: Mainly in lowlands, and most common in dry lowlands. Thinly distributed in mid-hills.

Voice: At times utters a series of repeated, loud, monotonous, metallic *chak* notes. Also utters some harsh, quarrelsome-sounding notes, especially when interacting with another roller. Juveniles make a harsh *kaaa* reminiscent of a violently angry cat.

Status: Resident.

At rest, brown mantle feathers can give a dull appearance when viewed from back. Strikingly blue-banded upperwings in flight.

Dollarbird *Eurystomus orientalis*

Size: 31cm

Habitat: Tall primary forests and mature secondary rainforests.

Distribution: Disjunctive distribution with records in South-west from Sinharaja and Kithulgala; earlier records from tall forests in South-east, and other records from North-east.

Voice: Nasal, sharp, repeated *krak*, sometimes repeated rapidly.

Status: Highly Scarce Resident.

Stout red bill, red legs and overall dark blue (almost black in poor light) separate this from the other two species of roller. Gets its name from 'dollars' or 'silverbacks' on underside of middle of primaries seen in flight.

Hoopoes Family *Upupidae*
Hoopoes have thin, long bills, rounded wings and jaunty crests. The family has just one species, though some authors split the species found in Africa.

Common Hoopoe
Upupa epops ceylonensis

Size: 31cm

Habitat: Uncommon bird best encountered in scrub forests in national parks in dry zone.

Distribution: Throughout lowlands, especially in dry zone.

Voice: Call is an onomatopoeic *hoopoe*, but most often sounds like an *oop oop*.

Status: Uncommon Resident.

Striking bird with wings and tail boldly barred in black and white. Long, pointed bill slightly downcurved. Body pinkish-cinnamon and head with a crest.

> **Hornbills** Family *Bucerotidae*
> Large birds with very big bills filled with a
> honeycomb structure. Not all species have a
> 'casque' on bill. Males imprison females in nest
> holes. Loud calls.

CEYLON GREY HORNBILL

Ocyceros gingalensis **e**

Size: 59cm

Habitat: Found in almost every sizeable forest patch in lowlands and hills.

Distribution: Widespread and occurs in all but the highest mountains.

Voice: Utters a juddering call that rolls for a few seconds. Also has a harsh, far-carrying *kraaa* contact call.

Status: Endemic. The more widespread of the two hornbill species found in Sri Lanka.

Female has dark mandibles with yellow crescent-like 'island' on upper mandible. Male has yellow mandibles with dark patch at base of upper mandible. Immature has all-yellow bill. Overall grey upperparts and lack of casque on upper mandible make confusion unlikely with larger Malabar Pied Hornbill (see right).

MALABAR PIED HORNBILL

Anthracoceros coronatus

Male (left) Female (right)

Size: 92cm

Habitat: Areas of large forest with old trees in dry lowlands.

Distribution: Widespread in dry lowlands. Absent from Northern Peninsula. Mainly frugivorous, but also eats small mammals and invertebrates when it can. Dependent on old forests for nest sites.

Voice: Call is raucous and far carrying. A single bird calling gives the impression of a flock of birds engaged in a squabble.

Status: Resident.

Large black-and-white bird with enormous bill. Female has a pale area around eye and lacks dark lining on rear of casque. Both sexes have a black-and-pale area at base of lower mandible.

PICIFORMES
Of the six families in this order, two are found in Sri Lanka, the barbets and woodpeckers, both adapted to an arboreal life. Woodpeckers are the most species rich of the families in the order.

Barbets Family *Capitonidae*
Barbets are compact, arboreal birds with stout bills used for excavating nests in tree trunks. They have bristles around the bill, and utter far-carrying calls from treetops. Though they are typically seen as fruit eaters, most species are omnivorous.

BROWN-HEADED BARBET *Megalaima zeylanica zeylanica*

Size: 27cm

Habitat: Well-wooded gardens and village garden habitats in lowlands and hills.

Distribution: From lowlands to higher hills; absent only in the highest mountains,

Voice: Varied. Most regular call is a repeated *turrok turrok*, which can start with a roll. Also has a nasal call that is reminiscent of that of a Black-headed Oriole.

Status: Common Resident.

Largest of the barbets in Sri Lanka; brown head, orange-red bill and facial patch, and green body. Confusion unlikely with any of the other species.

YELLOW-FRONTED BARBET
Megalaima flavifrons ℮

Size: 21cm

Habitat: Lowlands and hills. In heavy forest, displaces Brown-headed Barbet (see left).

Distribution: Mainly in wet zone from lowlands to highlands. In lowlands prefers areas with good forest and those with a hilly aspect. The most common barbet in highlands.

Voice: Utters two calls regularly. One is a tremulous *trokur trokur* that is repeated. The other begins with a roll (similar to one of the calls of Brown-headed Barbet), and continues as a repeated monosyllabic *whurr whurr whurr*. The latter is one of the natural sounds I associate with highlands.

Status: Endemic.

As in all local barbets, body is green. Yellow forehead and blue on face. Confusion with other species is unlikely.

CEYLON SMALL BARBET
Megalaima rubricapillus ℮

Size: 17cm

Habitat: Wooded gardens and forests in lowlands and hills.

Distribution: From lowlands to mid-hills. Most common in wet zone. In dry lowlands may be found in riverine forests and areas with good stands of fruiting trees.

Voice: Repeated *op op op*. Coppersmith has a *tonk tonk tonk*. Both species can occur in the same area, and some practice is needed to distinguish their calls.

Status: Common Endemic.

The easiest way to tell this species apart from Coppersmith Barbet (see right) is from the call (see below). Superficially similar to Coppersmith, but latter has yellow underparts streaked with dark green. Ceylon Small Barbet has green underparts lightly streaked with yellow. Both species have a red crown and red necklace. Coppersmith has yellow on face bordered with dark blue. In Ceylon Small, face is orange with a bright, light blue border.

COPPERSMITH BARBET
Megalaima haemacephala indica

Size: 17cm

Habitat: Forested areas in dry lowlands. Occasionally in hills.

Distribution: Mainly in dry lowlands, and in lower hills in drier northern and eastern aspects of central mountains.

Voice: Monotonous *tonk tonk* as though a copper anvil is being struck by a blacksmith.

Status: Resident.

This species can be separated from Ceylon Small Barbet (see left) by lemon-yellow not orange-yellow patches around eyes, and green-streaked, pale yellowish underparts.

EURASIAN WRYNECK *Jynx torquilla*

Size: 19cm

Habitat: Open country with wooded copses or scrub. Preferred prey is ants.

Distribution: Can turn up anywhere in suitable habitat, with records ranging from Yala in dry lowlands to Horton Plains in highlands.

Voice: Alarm call is a series of hard *teck* notes. Song is a sequence of whinnying *keek-keek-keek* notes.

Status: Vagrant.

Short-billed and long-tailed bird with intricately patterned wings. Unlike any other species in woodpecker family. Underparts pale and barred with black. Scapulars are black, and with another central broad, dark line along the back, a three-stripe appearance is formed on the back. Descends to the ground to search for ants.

Woodpeckers Family *Picidae*
Most woodpecker species use their long bills to probe for insects and other invertebrates on tree trunks, though a few forage on the ground. They have loud, whinnying calls. Most species excavate nest holes in trees, and most are sexually dimorphic. Males drum or call from prominent perches. The birds have zygodactyl feet.

Female   *Male*

BROWN-CAPPED PYGMY WOODPECKER
Dendrocopos nanus

Size: 13cm

Habitat: Needs wooded areas, but can tolerate disturbed habitats; up to the 2000s was even found in outskirts of Colombo. Visits village gardens.

Distribution: Found throughout Sri Lanka.

Voice: Rapidly uttered *chip-chip-chip*, with the notes blurring into each other and almost becoming a trill.

Status: Uncommon Resident.

Smallest of woodpeckers in Sri Lanka, the size of a sparrow. Wings, mantle and tail are barred black and white. Black tail has white spots. Brown cap. Thick brown line from eye to hindneck, which bisects white face. Male has a little red mark under cap, absent in female.

YELLOW-FRONTED PIED WOODPECKER
Dendrocopos mahrattensis mahrattensis

Size: 18cm

Habitat: Mainly in scrub forests of dry lowlands.

Distribution: Confined to dry zone in lowlands.

Voice: Metallic, slightly tremulous *tleep tleep.*

Status: Uncommon Resident. Vulnerable on IUCN Red List 2007.

Male

Overall black-and-white barred, small woodpecker. May be confused with diminutive Brown-capped Pygmy Woodpecker (see p. 189), which is much smaller and has a broad dark stripe from eye to shoulder. Male has yellow crown with red at rear. Female's crown is all yellow without any red.

Male

RUFOUS WOODPECKER
Micropternus brachyurus jerdonii

Size: 25cm

Habitat: In dense forests in lowlands. Nests in nests of *Crematogaster* ants.

Distribution: Widespread from lowlands to mid-hills where good forest habitat remains. Absent in arid Northern Peninsula. In the early 1980s, I saw one in Horton Place in central Colombo visiting a stand of old trees. However, patches of suitable habitat for birds passing through are now absent in Colombo, and even then this was a remarkably unusual sighting.

Voice: Strident and shrill *kee kee* call.

Status: Uncommon Resident.

Overall impression is of a brown woodpecker. Can look quite dark in the densely shaded forests it is found in. On close inspection, fine black barring on wings and tail can be seen. Male has short red stripe on cheek.

LESSER YELLOWNAPE
Picus chlorolophus wellsi

Size: 27cm

Habitat: Well-wooded gardens and forests in lowlands and hills. Found in village gardens only where suitable forest patches occur nearby.

Distribution: Confined to wet zone in lowlands mid-hills.

Voice: Shrill, wheezy, whistled *wheeuw.*

Status: Resident.

Male

Female

Female

Green woodpecker with yellow nape in both sexes. Confusion possible with Streak-throated Woodpecker (see right), which lacks yellow nape and is streaked rather than barred underneath. Both sexes have red crowns, but female lacks red moustachial stripe of male. Note that it is unusual in woodpeckers for both sexes to have red on crown.

STREAK-THROATED WOODPECKER
Picus xanthopygaeus

Size: 29cm

Habitat: Visits well-wooded gardens in hills, sometimes descending to the ground to search for ants.

Distribution: An unusually restricted distribution in drier eastern aspects of highlands, descending to dry lowlands.

Voice: Contact call is a nasal *kik.*

Status: Scarce Resident.

Body grey-green with green wings and yellow rump. Scaly underparts. Both sexes have clear white supercilium. In male, lores are black and crown is red. In female, lores and crown are dark. Streak-throated lacks red moustachial stripe of Lesser Yellownape (see left). Confusion between the two species is unlikely.

BLACK-RUMPED FLAMEBACK
Dinopium benghalense

Size: 29cm

Habitat: Often encountered in coconut plantations.

Distribution: Black-rumped Flameback replaces Ceylon Red-backed Woodpecker in northern parts of Sri Lanka from around Puttalam to Jaffna Peninsula. Black-rumped Flameback has a very restricted distribution, whereas Ceylon Red-backed Woodpecker is widespread elsewhere. There is a narrow hybridization zone where the two species meet.

Voice: Short, sharp, whinnying scream.

Status: Uncommon Resident.

Male

Female has a black forehead flecked with white and a red crest. Male has a red forehead and crest. Black-rumped Flameback and Ceylon Red-backed Woodpecker (see right) are easily told apart from golden and red upperparts respectively. Until 2014, they were treated as 'red-backed' and 'golden-backed' subspecies of the same woodpecker.

CEYLON RED-BACKED WOODPECKER
Dinopium psarodes **e**

Male

Size: 29cm

Habitat: Gardens and well-wooded areas in lowlands and hills.

Distribution: Widespread throughout Sri Lanka, covering areas where Black-rumped Flameback is absent, but both species can overlap in range and hybridized forms occur.

Voice: Short, sharp, whinnying scream.

Status: Common Endemic.

Forehead black in female, red in male. Both sexes have red on hindcrown. Black-rumped Flameback (see left) was treated as having four subspecies on the Asian mainland and two subspecies in Sri Lanka. Puzzlingly, one of the races in Sri Lanka, race *psarodes*, had a red back. Since 2014, southern race *psarodes* has been recognized as a different species, the Ceylon Red-backed Woodpecker. 'Golden-backed' race *jaffnense* is now the sole Sri Lankan subspecies of Black-rumped Flameback.

CRIMSON-BACKED FLAMEBACK
Chrysocolaptes stricklandi ⓔ

Size: 33cm

Habitat: Well-wooded gardens and forests in lowlands and hills. Displaces Ceylon Red-backed Woodpecker in heavily forested patches.

Distribution: Absent in arid zones in North-west, North and South-east. Otherwise found in forested areas up to highest mountains. Prefers forests and absent from urban habitats in which Ceylon Red-backed Woodpecker is prevalent.

Voice: Call a repeated whinnying *tree tree tree tree*. Does not have the raucous, ringing insistence of Ceylon Red-backed Woodpecker's urgent-sounding call. Shrill, whinnying call also used in flight.

Status: Endemic.

Male

Female's crown is black with white flecks. More common Ceylon Red-backed Woodpecker (see p. 192) is similar, but Crimson-backed Flameback has an ivory-coloured bill. Crimson-backed Flameback also has a more complex facial pattern, with a white 'island' framed by two thin black moustachial stripes.

Male

WHITE-NAPED FLAMEBACK
Chrysocolaptes festivus tantus

Size: 29cm

Habitat: Found in wooded areas in dry lowlands.

Distribution: Mainly in North Central Province and South-east from around Hambantota to Yala.

Voice: Contact calls are tremulous, very reminiscent of calls of Ceylon Red-backed Woodpecker (see p. 192), but not as loud and grating. They are much shorter in duration.

Status: Scarce Resident. Vulnerable on IUCN Red List.

Forecrown of female is speckled black and white, with a yellow crown.

PASSERIFORMES
Sixty per cent of the birds in the world are in this order, comprising 38 families. They are diverse in size and behaviour.

Pittas Family *Pittidae*
Pittas are beautiful ground-dwelling species favouring shaded undergrowth.

INDIAN PITTA
Pitta brachyura brachyura

Size: 19cm

Habitat: Gardens and forests. Favours dense undergrowth, where it can be seen hopping around. Will fly up and perch in mid-canopy, from where it may call.

Distribution: Spreads widely throughout Sri Lanka on arrival. Large numbers of this migrant must arrive, as it is not unusual even in Colombo for people to find exhausted pittas on arrival. On an acre of land in Talangama Wetland, I have noticed three Indian Pittas taking up residence in some years.

Voice: Local Sinhala name *Avichchiya* is onomatopoeic. It is also called the Six o'clock Bird as it calls predictably at this time at dusk. Also makes a series of harsh, scolding calls.

Status: Migrant.

Green upperparts, red vent and bold black eye-stripe. In flight, bases of primaries have a prominent white flash. The only pitta in Sri Lanka. Typical dumpy profile of a pitta.

Larks Family *Alaudidae*
These insectivorous birds are usually sandy-brown and favour grassland.

JERDON'S BUSHLARK
Mirafra affinis ceylonensis

Size: 14cm

Habitat: Open areas in dry lowlands.

Distribution: Dry lowlands.

Voice: Series of thin, wispy, ascending *tseep tseep* notes. The bird ascends while singing, then makes a 'parachute descent' in courtship.

Status: Resident.

Similar to Oriental Skylark (see p. 196). Shorter, chunky bill helps to distinguish it from slimmer, longer billed pipits. At times, face is chestnut bordered by clear white supercilium on top and white margin at back of cheek. Most of the time cheek is streaked. Shorter tail also separates it from pipits. Rufous on wing does not show well at rest.

Male

Female

ASHY-CROWNED FINCH-LARK
Eremopterix griseus

Size: 13cm

Habitat: Open areas in dry lowlands. Almost entirely ground dwelling.

Distribution: Present in dry lowlands. In December 2003, I was one of many observers to photograph a tired female in Horton Plains National Park, possibly one from a flock that was travelling. This could have been a migrant bird from Asia that was flying over the mountains to reach the South (I suspect many birds considered to be resident species are actually supplemented by wintering populations).

Voice: Song has long, trilling notes like a fishing line being drawn in. Also some drawn-out, single *whee* notes.

Status: Uncommon Resident.

Female is a drab, sparrow-like bird, though it lacks chestnut flight feathers of female House Sparrow (see p. 260), and has a weaker bill. Males have blackish throats. Males in breeding plumage are striking, with black underparts, white face and black line from base of bill to neck.

GREATER SHORT-TOED LARK
Calandrella brachydactyla longipennis

Size: 15cm

Habitat: Open, stony grassland and arid country. On migration may turn up in wet grassland.

Distribution: May turn up anywhere as it can winter in arid open country as well as in wet grassland.

Voice: Musical chitter, reminiscent of a skylark but very different.

Status: Vagrant.

Five species of *Calandrella* larks have been recorded on the Indian subcontinent as migrants. Careful notes need to be taken of details of such birds, for instance whether the primaries extend beyond the tertials; if there streaking on the face, a dark side-patch on the sides of the breast near the wing, or streaking on the breast and whether this is fine. Greater Short-toed Lark has a dark breast side-patch, and an unstreaked face and underparts; primaries do not extend beyond tertials. Also has a stout bill, prominent white eye-stripe and eye-ring. Upperparts warm with conspicuous streaking.

ORIENTAL SKYLARK *Alauda gulgula*

Size: 16cm

Habitat: Open areas in dry lowlands and hills. Most prevalent in arid habitats. The skies are full of this skylark's song in the area around Mannar.

Distribution: Found in arid South-east and extends along east coast to dry lowlands from north-central plains to Northern Peninsula.

Voice: Series of *chee chee* notes interspersed with notes of different pitch, and trills and chirps. There is a pattern to the sequence, but sequence is complex and keeps changing. Singing bird usually ascends the sky as it sings and is easily located.

Status: Two resident races described in Sri Lanka. Race *gulgula* is an Uncommon Resident in dry lowlands, and race *australis* is a Scarce Resident in higher hills.

Both Jerdon's Bushlark (see p. 194) and Oriental Skylark have well-marked ear-coverts. Jerdon's Bushlark is also known as Rufous-winged Bushlark, but does not show rufous flight feathers at rest. Confusion is likely between the two species at rest. Oriental Skylark has a finer bill than Jerdon's Bushlark. It is chunkier in build than Paddyfield Pipit (see p. 204), with which confusion is possible. Oriental Skylark often raises a crest of feathers on its crown.

Swallows and martins Family *Hirundinidae*
These long-winged, fast-flying, insectivorous birds catch their prey on the wing. They are unrelated to the similar looking swifts, which are in a different scientific order. The similarity of the swallows and martins to the unrelated swifts is an example of convergent evolution.

COMMON SAND-MARTIN
Riparia riparia

Size: 13cm

Habitat: In its breeding range, found near water where there are suitable earthen banks for nesting. Generally found hawking for insects over areas with water and fields, or patchworks of wood and reed beds or grassland.

Distribution: May turn up anywhere, though most records in Sri Lanka have been in the South-east, where large flocks of Barn Swallows congregate in winter.

Voice: Tremulous *kreek-kreek* that is repeated, occasionally with slight variations.

Status: Scarce Migrant.

White underparts and sandy-brown upperparts. Sandy-brown breast-band can be variable in size. Brown extends from crown to over eye and cheeks. The only other brown hirundine in Sri Lanka is the Vagrant Dusky Crag-martin (see p. 197), which is uniformly overall a darker brown, with a row of white spots on tail. Juvenile Common Sand-martin has pale fringes on coverts and scapulars.

DUSKY CRAG-MARTIN
Ptyonoprogne concolor

Size: 13cm

Habitat: Breeds in mountains and gorges, but may also nest in old buildings.

Distribution: May turn up anywhere.

Voice: Repeated *tchree-tchree*, with an audible gap between the two notes. A little reedy.

Status: Vagrant.

The only all-brown hirundine recorded in Sri Lanka. Tail has row of white spots. Altogether six species of brown martin breed on the Indian subcontinent and may turn up as vagrants in Sri Lanka. For any martin, note extent of white on underparts, the presence of any scaling on vent, tail-spots and how dark or pale it is.

Juvenile

BARN SWALLOW *Hirundo rustica*

Size: 18cm

Habitat: Hunts over open areas, using tall trees or telephone lines as a perch.

Distribution: Spreads throughout Sri Lanka up to highlands. Before departure, large flocks gather in lowlands; especially visible in southern coastal strip near Bundala National Park.

Voice: Metallic chirping notes with a trill at the end of the sequence that is repeated.

Status: Race *gutturalis* is a Common Migrant, *rustica* is an Uncommon Migrant and *tytleri* is a Scarce Migrant.

European race *rustica* has complete black breast-band. Asian race *gutturalis* has incomplete black breast-band and is more common. Race *tytleri*, also known as Tytler's Swallow, is suffused with pink on underparts, but some individuals in the other two races can also be tinged pink on underparts.

HILL SWALLOW *Hirundo domicola*

Size: 13cm

Habitat: Hills and highlands. Will nest in buildings.

Distribution: Mainly in wet zone up to highlands. Becomes more common with increasing elevation.

Voice: Soft, chittering calls at rest and in flight.

Status: Resident.

Juvenile

'Dirty white' underparts. No black breast-band. Tail forked as in migrant Barn Swallow (see p. 197), but lacks long outer tail-streamers.

WIRE-TAILED SWALLOW
Hirundo smithii

Size: 14cm

Habitat: Found in open country and cultivated land near water.

Distribution: May turn up anywhere. First record was in dry lowlands in the North-west.

Voice: Call is a repeated, sharp *tchik.*

Status: Vagrant.

Red cap, dark mask on face widening out towards nape, glossy blue upperparts and clean white underparts help to identify this species. Adults have long, filamentous projections to outer-tail feathers. Sometimes these are hard to see or are absent because they have broken.

RED-RUMPED SWALLOW
Hirundo daurica

Size: 19cm

Habitat: Open areas interspersed with forest in dry lowlands.

Distribution: Recorded in dry lowlands. Can occur in large flocks. Uda Walawe National Park is one of the best sites in which to see it.

Voice: Sounds similar in pattern to Ceylon Swallow's, but is less liquid in quality.

Status: Scarce Migrant.

Race *erythropygia* is the Indian Red-rumped Swallow, one of several races of this widespread swallow. Indian race has finer streaking than Nepali race *nipalensis*, a Highly Scarce Migrant. Underparts much paler than those of Ceylon Swallow (see right), which was once considered another race of Red-rumped Swallow *H. daurica*. Rump can also look very pale in flight.

CEYLON SWALLOW
Hirundo hyperythra **e**

Size: 18cm

Habitat: Open habitats such as grassland, coastal areas, paddy fields and similar places.

Distribution: Found throughout Sri Lanka.

Voice: Call, often uttered in flight, has a liquid, bubbling quality to it.

Status: Endemic.

Deep red underparts and rump help distinguish this species from migrant races of Red-rumped Swallow (see left).

STREAK-THROATED SWALLOW

Hirundo fluvicola

Size: 12cm

Habitat: Open country near water.

Distribution: First record was in highlands, but could turn up anywhere.

Voice: Rolling *brrt*.

Status: Vagrant.

Red cap, blue mantle and white underparts provide the same basic design as that of Wire-tailed Swallow (see p. 198), but Streak-throated Swallow is easily distinguished from Wire-tailed due to strong brown streaking on face, throat and breast, eye not covered by a mask, blue on upperparts confined to mantle, and brown wings and rump. Square-ended tail with no streamers.

Pipits and wagtails Family *Motacillidae*
These are long-tailed birds of grassland and marshes with a habit of wagging the tail. Pipits are cryptically coloured. Wagtails show pronounced differences between breeding and non-breeding plumages.

FOREST WAGTAIL *Dendronanthus indicus*

Size: 17cm

Habitat: Forest patches or large home gardens. Forages in leaf litter in densely shaded thickets.

Distribution: On arrival spreads throughout Sri Lanka.

Voice: Characteristic *plink plink* call that betrays its presence.

Status: Migrant.

Unlikely to be confused with any of the other wagtails, with its boldly white-barred, blackish wings, brown mantle and crown, and white underparts. Two broad black bands on breast. Both sexes have the two bands. Sexes cannot be told apart.

White Wagtail *Motacilla alba*

Size: 18cm

Habitat: Found near waterbodies on adjoining damp grassland.

Distribution: May turn up anywhere in suitable habitats. Periodic visitor to the spill at Uda Walawe National Park.

Voice: Rapid *chee-chee* is contact call, and may be mixed with warbling notes. Song is a series of repeated *chizwee* notes.

Status: Race *dukhunensis* is a Scarce Migrant, and race *leucopsis* is a Highly Scarce Migrant.

White forehead and black rear crown in adults. In male, black has sharp borders with grey back. In breeding plumage chin and throat are black. This fades away to white with an uneven edge to remaining black on breast. First-winter bird has a dark breast-band and grey crown. Wings mainly black, edged with white. No yellow in any stage.

Juvenile

Summer, male

Winter

White-browed Wagtail
Motacilla maderaspatensis

Size: 21cm

Habitat: Grassy pastures.

Distribution: Recorded in mid-hills around Kandy, but could arrive anywhere in Sri Lanka where there is suitable habitat.

Voice: Repeated, triple-noted *bruh-bur-cheeez*, interspersed with other short notes. Some notes reminiscent of flight call of Grey Wagtail (see p. 203). Also some quick *cheez-cheez* notes.

Status: Vagrant.

Distinct white supercilium against all-black head makes confusion unlikely with different races of White Wagtail M. *alba*. Black extends from back to upper tail.

CITRINE WAGTAIL
Motacilla citreola citreola

Size: 17cm

Habitat: Wet meadows and banks of watercourses.

Distribution: Recorded in dry lowlands in locations such as Uda Walawe National Park.

Voice: Mix of repeated *tseep* notes, with a *chizwith* or *tschirrup* thrown in.

Status: Highly Scarce Migrant.

Winter

Winter

In breeding plumage, yellow head and body with grey mantle, and black-and-white wings with two wing-bars. Black edge from bend of wing along where nape meets mantle. In winter black edging is lost and mantle is less grey, but wings retain two white wing-bars. Underparts are not bright yellow, but head and breast remain tinged with yellow. Crown and cheek-patch greenish-brown, and lacks dazzling yellow head and black bill of breeding bird. First-winter bird lacks yellow wing markings and has facial markings similar to adult winter. Confusion with breeding Yellow-headed Wagtail M. *flava lutea* is possible, as it lacks a grey back (yellow to greenish-yellow) and lacks black edging. Race *citreola* recorded in Sri Lanka has a greyish-brown mantle contrasting with a yellow face and underparts in adult. Juveniles and first-winter birds lack yellow and can show a lot of brown with white wing-bars. Brownish ear-coverts bordered by pale crescent that curves behind eye. This helps to distinguish immatures from immature Western Yellow and Grey Wagtails (see p. 203).

Telling apart Western Yellow Wagtails in Breeding Plumage

	thunbergi **Grey-headed**	*feldegg* **Black-headed**	*beema*	*lutea*
Crown	Grey	Black	Grey	Yellow
Cheeks and ear-coverts	Black, contrasts with crown	Black, concolorous with crown	Pale, bordered with grey	Yellow
Eye-stripe (supercilium)	NA	NA	White	Yellow, and may show little contrast with crown

Western Yellow Wagtail
Motacilla flava

Size: 17cm

Habitat: Wet habitats in lowlands. Lake edges, marshes, ploughed paddy fields and similar places.

Distribution: Spreads throughout Sri Lanka on arrival.

Voice: High-pitched *shree shree* double-noted call. Very different from single, repeated note of Grey Wagtail (see right).

Status: Race *thunbergi* is a regular Migrant in small numbers; the other migrant races are rare.

Advanced field guides may need to be consulted to distinguish the several races that occur (or potentially occur) in Sri Lanka. The table opposite flags some of the key features of birds in breeding plumage. Adults in non-breeding plumage and immatures can present significant identification challenges. By and large, most of the yellow wagtails seen are likely to be of race *thunbergi*, or Grey-headed Yellow Wagtail.

Winter

Winter

Winter

Winter

Grey Wagtail
Motacilla cinerea melanope

Size: 17cm

Habitat: Favours banks and grassy areas adjacent to water.

Distribution: Spreads throughout Sri Lanka on arrival. Seems to prefer high elevations, and the likelihood of encountering it increases with elevation.

Voice: Clearly articulated *tchink tchink* uttered in quick succession and repeated.

Status: Migrant.

Grey crown and mantle, white supercilium and yellow underparts with some white on flanks. Non-breeding adult has white throat. Breeding adult has black chin and throat, and white malar stripe. Most birds seen in Sri Lanka are in non-breeding plumage.

RICHARD'S PIPIT *Anthus richardi richardi*

Size: 18cm

Habitat: Short grassland and open areas in dry lowlands.

Distribution: Appears more abundant in northern parts of Sri Lanka.

Voice: A distinct, repeated *cheep*. Quite different from Paddyfield Pipit's voice.

Status: Scarce Migrant.

Pale lores (can look dark at times), usually has an upright stance and gives an impression of having to drag a heavy tail. Larger than Paddyfield Pipit (see right), but this may not be apparent unless a direct comparison is possible. Warm buff flanks and long hind claw.

PADDYFIELD PIPIT
Anthus rufulus malayensis

Size: 15cm

Habitat: Areas of short grass.

Distribution: Widespread from lowlands to highlands.

Voice: Call sequence is *chip chip chip-chip-chip*.

Status: Resident.

Common pipit; for identification of pipits, it is best to learn this species and use it as a baseline. Unfortunately, there is no clear feature that makes a bird a Paddyfield Pipit, and the identification is relative. Confusion is most likely with the scarce migrant Richard's Pipit (see left). Like Richard's, it has tawny flanks and a long hindclaw, but the bill is finer. Long hindclaw is not always apparent in the field. Differences in calls provide the best means to separate the species. Paddyfield Pipits can look 'different' due to age, moult or lighting conditions. At times, they can raise the feathers on the crown. Supercilium and malar stripe are distinct and cheek-coverts are rusty, but these features are shared with other pipits.

TAWNY PIPIT *Anthus campestris*

Size: 15cm

Habitat: Sandy dunes and barren mountain slopes.

Distribution: Most likely to be recorded in dry lowlands.

Voice: Loud *chillip* or *chep*. Song a repeated trilling *chooroo-wee*.

Status: Vagrant.

Overall plain-looking pipit identified by dark lores, fine bill, horizontal, wagtail-like stance and call. Adults and first-winter birds may have fine streaking on upperparts. Juveniles have heavier streaking.

BLYTH'S PIPIT *Anthus godlewskii*

Size: 17cm

Habitat: Grassy pastures in dry lowlands.

Distribution: Recorded mainly in Uda Walawe National Park, where it is a recurrent visitor.

Voice: Repeated, nasal, single-noted *chew*.

Status: Scarce Migrant.

'Blunt-tipped' median coverts and call are key diagnostics. Streaked bands on upper breast seem finer and tidier compared with Paddyfield Pipit's (see p. 204). Blyth's and Richard's Pipits (see p. 204) all have an unstreaked belly. Compared with Richard's, bill in Blyth's is finer.

Olive-backed Pipit
Anthus hodgsoni hodgsoni

Size: 15cm

Habitat: Mixture of open habitats, including grassland set in wooded areas.

Distribution: May turn up anywhere, but lowlands are most likely.

Voice: Flight call is a drawn-out, thin *tseez*. Song is complex and includes sharp notes and warbles. Reminiscent of some European *Sylvia* warblers.

Status: Vagrant.

Superficially similar to first-winter Red-throated Pipit (see right), as facial patterns are similar, with thick white supercilium behind eye, buffy supercilium in front of eye, well-defined brown ear-coverts, and black malar stripe joining pronounced and continuous flank-streaks on white underparts. Mantle of Olive-backed Pipit is greenish and has only lightly defined streaks. Red-throated adults and first-winter birds show heavy black streaking on a brown mantle.

Red-throated Pipit
Anthus cervinus

Size: 15cm

Habitat: Marshes, grassland and cultivated land near water in winter quarters. In breeding range found in swampland as well as upland birch forest with grassland.

Distribution: May arrive anywhere, but most likely in lowlands.

Voice: High-pitched *psee* that fades away. Distinctive call. Song is complex and usually includes a series of ascending, repeated *zee-zee* or *chwee-chwee* notes, which are mixed with warbles and other repeated double notes.

Status: Vagrant.

Breeding adults have reddish faces, and red on chin and throat down to breast. Unmistakable in breeding plumage. First-winter birds and adults in non-breeding plumage have heavy streaking on a brown mantle. This separates them from Olive-backed Pipit, which has ill-defined streaks on an olive back (see also Olive-backed Pipit, left).

> **Cuckooshrikes** Family *Campephagidae*
> This varied family includes the cuckooshrikes, the
> flycatcher-shrikes and the colourful minivets.

LARGE CUCKOOSHRIKE
Coracina macei layardi

Size: 28cm

Habitat: Well-wooded areas.

Distribution: Widespread except in highlands. Though primarily a forest bird, it can tolerate degraded habitats and may be seen close to towns.

Voice: Loud, nasal, repeated drawn-out note.

Status: Uncommon Resident.

Female

Female distinctly barred on underparts; male has faint barring. Females of Large Cuckooshrike and Black-headed Cuckooshrike (see right) can be told apart by larger size and dark mask over eye of Large. It also has a heavier bill and chunkier build.

Male   *Female*

BLACK-HEADED CUCKOOSHRIKE
Coracina melanoptera sykesi

Size: 20cm

Habitat: Wooded patches and gardens throughout lowlands and hills.

Distribution: Widespread except in highlands.

Voice: Beautiful musical *yip yip yip yee yee yee* notes.

Status: Resident.

Female lacks black head and is white below with fine grey barring. Black-headed Cuckooshrike male and female have white vent and undertail-coverts, lacking faint barring seen in Large Cuckooshrike (see left). Both sexes have dark flight feathers; they are black in male, duller in female.

SMALL MINIVET
Pericrocotus cinnamomeus cinnamomeus

Size: 15cm

Habitat: Wooded patches. Partial to riverine vegetation.

Distribution: Widespread throughout Sri Lanka.

Voice: Thin, wispy, high-pitched calls with many notes in sequence that vary in pitch.

Status: Resident.

Both sexes have orange rumps, sides of tail and wing-patch. In female, orange areas are more yellowish than in male. Entire head of male is dark grey with a colourful orange breast shading into a pale vent. Throat of female has white shading into yellow on underparts. Often in small flocks.

Male

Female

Female   *Male*

ORANGE MINIVET
Pericrocotus flammeus flammeus

Size: 19cm

Habitat: Wooded patches.

Distribution: Widespread throughout Sri Lanka.

Voice: Explosive *zeet zeet* calls herald arrival of a flock. Complex sequence of whistled notes.

Status: Resident.

Male glossy black on head and upperparts with large flash of scarlet on black wings. Confusion unlikely with male Small Minivet (see left), which is grey rather than glossy black. In female Orange black is replaced with grey, and scarlet is replaced with yellow. Throat is yellow. Almost always in small flocks that betray their presence with calls as they fly about. Prefers to feed in the canopy. The birds seem restless and flocks seem to be always on the move.

Pied Flycatcher-shrike
Hemipus picatus leggei

Size: 14cm

Habitat: Well-wooded areas. Visits degraded habitats if they adjoin good patches of forest.

Distribution: Throughout Sri Lanka.

Voice: Repeated pair of trilling notes of medium length. Trilling call is distinctive and unlikely to be confused with that of other species in Sri Lanka.

Status: Uncommon Resident.

Ceylon Woodshrike
Tephrodornis affinis ⓔ

Size: 18cm

Habitat: Scrub jungle in dry lowlands. Occupies low canopy and is not difficult to see on mature trees around waterholes in Yala. Feeds on insects.

Distribution: Mainly in dry lowlands, but there is a strip in wet zone encompassing Meetirigala Forest where it is present. Distribution pattern is curious.

Voice: Has 2–3 regular calls. The most frequent is a descending *whee whee whee* of 4–5 notes. Another is a lovely musical whistled call that ends with *peep peep peep*.

Status: Endemic.

Overall impression is of a small, black-and-white bird. Dirty white or greyish underparts with black upperparts. Black wings have prominent white wing-bar. Almost always in small flocks that prefer the canopy. Can be overlooked if attention is not paid to its calls.

Sexes are similar. Overall greyish or greyish-brown bird with a dark mask. Underparts are pale, mostly white. Blackish mask bordered by white supercilium above and below is another band of white that lightly contrasts with dirty white breast, which fades into white belly.

> **Monarchs** Family *Monarchidae*
> Monarchs are flycatchers with an active demeanour. They are prone to making short, sudden flights between trees as they forage. They do not return repeatedly to the same perch as typical flycatchers do.

ASIAN PARADISE FLYCATCHER (INDIAN PARADISE FLYCATCHER)
Terpsiphone paradisi paradisi

Size: 20cm/50cm w/streamers

Habitat: Lowlands and hills in forests and well-wooded gardens.

Distribution: Migrant race spreads all over the island from lowlands to highlands. Paradise flycatchers seen in cities such as Colombo in the West during migrant season are all of this race.

Voice: Harsh *krik* with a nasal intonation, repeated at a few seconds apart, comprises contact call. Song a beautiful, soft melody of various musical notes mixed with some rapid sequences.

Status: Migrant.

Male

Two races occur in Sri Lanka, resident race *ceylonensis* (described right) and migrant race *paradisi*. Fully mature adult males are easy to tell apart as Indian is black and white, while Ceylon has a black head, chestnut upperparts and white underparts. Indian males gradually acquire white colouration, from being chestnut in the first year. By the second year, migrant Indian males begin to acquire white feathers and show a mix of chestnut and white. By the third year, male plumage is completely white. Females of both races are similar and lack tail-streamers.

ASIAN PARADISE FLYCATCHER (CEYLON PARADISE FLYCATCHER)
Terpsiphone paradisi ceylonensis

Male

Female

Size: 20cm/50cm w/streamers

Habitat: Lowlands and hills in forests and well-wooded gardens.

Distribution: Dry lowlands.

Voice: Like Indian Paradise Flycatcher's.

Status: Resident.

In resident race *ceylonensis*, males are chestnut throughout their lives (see more details under Indian Paradise Flycatcher, above).

BLACK-NAPED BLUE MONARCH

Hypothymis azurea ceylonensis

Size: 16cm

Habitat: Forests.

Distribution: Throughout Sri Lanka up to mid-hills

Voice: Repeated, explosive *chwizz*.

Status: Uncommon Resident.

Habit of actively foraging in mid-canopy and not staying for long helps separate this bird from less active and sedentary feeding behaviour of Tickell's Blue Flycatcher (see p. 226), which it superficially resembles. Also makes sharp *zit* call as it forages. Tickell's Blue Flycatcher has orange-red breast compared with blue of Black-naped Blue Monarch. Juvenile monarchs have blue heads, but rest of plumage can look plain grey.

Fantails Family *Rhipiduridae*
Fantails are black-and-white birds with long tails that are fanned out regularly. They make 'drunken piper' movements as they fly here and there in a seemingly random series of zigzag movements, and have beautiful songs.

WHITE-BROWED FANTAIL

Rhipidura aureola compressirostris

Size: 18cm

Habitat: Forests. Visits gardens in highlands.

Distribution: Strangely absent in western lowlands, but occupies wet zone in highlands. Found throughout dry lowlands. Feeds on insects and other invertebrates, and flies onto house verandahs to pick up insects.

Voice: Range of melodious calls and plaintive, whistled song.

Status: Resident.

Black-and-white bird with relatively long 'fan tail'. Prominent white supercilium. The only fantail in Sri Lanka. Planters in highlands used to call the bird the Drunken Piper on account of its erratic behaviour, with swooping flights.

> **Bulbuls** Family *Pycnonotidae*
> These perky, small birds perch in a 'bolt-upright' position. Calls and songs of different species in the family are highly variable, so there is no discernible family quality to them.

BLACK-CAPPED BULBUL
Pycnonotus melanicterus 🇪

Size: 17cm

Habitat: Forests in lowlands and hills.

Distribution: Mainly in wet zone up to mid-hills. Also present in dry-zone forests, but in much lower numbers than in wet-zone forests.

Voice: Most frequent calls include sequence with *pip pip* notes; another sequence contains several churring notes like those of Grasshopper Warbler (see p. 242). Song has a series of rising melancholy syllables.

Status: Uncommon Endemic.

Black cap and yellow plumage are distinctive. In flight often shows white tips on tail feathers.

RED-VENTED BULBUL
Pycnonotus cafer haemorrhousus

Size: 20cm

Habitat: Variety of habitats from scrubland to home gardens. Even found in major cities such as Colombo.

Distribution: Widespread, and found throughout Sri Lanka. One of the most common birds in Sri Lanka.

Voice: Tremulous, two-note call that sounds like *teacher bread* and at times *pitta bread*. In the evenings the birds utter a churring noise.

Status: Common Resident.

Greyish-brown bulbul with pale edges to feathers giving a scalloped look. Head black; dusky breast shading to white ending in bright red vent. Tail tipped with white.

YELLOW-EARED BULBUL
Pycnonotus penicillatus Ⓔ

Size: 20cm

Habitat: Hills and highlands. Visits gardens in highlands.

Distribution: In wet zone from mid-hills to highlands. Becomes more common with increasing elevations.

Voice: Draws attention to itself with loud calls of *pik pik pik*. Call also uttered in flight. Also has a guttural call.

Status: Endemic. Vulnerable on IUCN Red List.

Face strikingly patterned in black and white with yellow ear-tufts. Inner webs of flight feathers are leaden in colour, and outer webs are green. Most of the time, only green shows and body looks olive-green and yellow.

WHITE-BROWED BULBUL
Pycnonotus luteolus insulae

Size: 20cm

Habitat: Forests and well-wooded gardens in lowlands and hills.

Distribution: Found throughout Sri Lanka except in highlands.

Voice: Uplifting *chik-chee-chee-chik-chik* followed immediately by two grating notes and musical chirps. Complex call sequence.

Status: Resident.

Nondescript brown bulbul with pale brown upperparts and dirty white underparts. White eyebrow is distinct most of the time. Yellow malar stripe and vent. Can be discreet in behaviour and presence is picked out by call.

YELLOW-BROWED BULBUL *Iole indica*

Size: 20cm

Habitat: Forest patches, mainly in wet lowlands.

Distribution: From lowlands to mid-hills. Two races, *guglielmi* and *indica*, described from wet zone and dry zone, but they are indistinguishable in the field. Dry zone race is scarce. Most likely to be seen in wet zone forests and adjoining village gardens.

Voice: Characteristic *quick quick* call.

Status: Uncommon Resident.

Overall yellow bulbul, more yellow on underparts and olive-yellow on upperparts. Yellow encircles eye and runs from base of bill to eye.

SQUARE-TAILED BLACK BULBUL
Hypsipetes ganeesa

Size: 23cm

Habitat: Forest patches.

Distribution: Mainly in wet lowlands, but locally found in dryzone riverine forests.

Voice: Loud, tremulous *chik-chuo*. Also series of wailing calls interspersed with indignant chattering notes.

Status: Uncommon Resident.

Black bulbul with long, pointed red bill and red legs. Frequents tree canopies, from which it calls noisily.

Ioras Family *Aegithinidae*
Ioras are arboreal, small yellow birds, usually with black and white on the wing.

COMMON IORA
Aegithina tiphia multicolor

Size: 14cm

Habitat: Forests and wooded gardens.

Distribution: Widespread from lowlands to mid-hills. Most numerous in dry lowlands.

Voice: Soulful, long-drawn out notes. These and those of Indian Black Robin (see p. 230) are two of the acoustic signatures of daytime dry lowlands on hot afternoons.

Status: Resident.

Male in breeding plumage acquires black upperparts and black tail. It is then distinctive from female, which lacks the black. In non-breeding plumage, male is similar to female, but panel on wings may show more black. Marshall's Iora (see right) is found in east. In Marshall's, both sexes have white outer tail feathers and much more white on wings. I once saw several ioras singing close to each other almost as if they were in a lek.

Male *Female*

Juvenile

MARSHALL'S IORA
Aegithina nigrolutea

Size: 14cm

Habitat: Scrub forest.

Distribution: Restricted to east and south-east of Sri Lanka, with records ranging from Buttala and Nilgala areas to Lunugamwehera, which is close to Yala National Park.

Voice: Contact call a harsh chatter that can end in a quivering repetition. Reminiscent of Common Iora's contact calls. Song includes some musical notes with a slight melancholy tone. Different from widespread and abundant Common Iora.

Status: Scarce Resident.

Both sexes have white on sides of tail and tips. Two white wing-bars, as in Common Iora (see left), but there is more white edging on flight feathers, and tertials are especially more marked with white edges. Probably overlooked in the past for Common Iora, and only recorded as a breeding species in Sri Lanka in March 2006 by Chinthaka Kaluthota after its re-discovery.

Breeding male

> **Leafbirds** Family *Chloropseidae*
> These are green birds of the canopy with variable and ventriloquistic calls. They are accomplished mimics, incorporating the vocalizations of other birds into their own.

GOLD-FRONTED LEAFBIRD
Chloropsis aurifrons insularis

Size: 19cm

Habitat: Throughout lowlands and hills in wooded gardens and forests.

Distribution: Widespread except in highlands.

Voice: Complex series of notes as in Jerdon's Leafbird, but plainer and less trilling. With practice, the two species can be told apart by ear.

Status: Uncommon Resident. Less common than Jerdon's Leafbird.

Male

Male can look similar to Jerdon's Leafbird (see right), but black edge to blue throat gorget does not have diffused brownish-yellow between it and green of body. Female's forecrown is yellowish and lacks male's golden colour. Juvenile is all green.

JERDON'S LEAFBIRD
Chloropsis jerdoni

Male

Size: 19cm

Habitat: Throughout lowlands and hills in wooded gardens and forests.

Distribution: Widespread except in highlands.

Voice: Very complex sequence of notes, often borrowing elements of calls and songs from other birds. Typically consists of whistled notes, trills and chirps. Throws in almost every type of call a passerine is capable of uttering.

Status: Resident.

Male's facial pattern similar to that of Gold-fronted Leafbird (see left), but black surrounding purple centre is not as wide and extensive. Male's crown is green. Female has pale blue throat faintly edged with yellow. Juvenile is similar.

Shrikes Family *Laniidae*
Shrikes are long-tailed, carnivorous birds with hook-tipped bills and a reputation for impaling their prey on thorns to build a larder. This habit is rarely seen in Sri Lanka. The only resident species is confined to the arid North.

BROWN SHRIKE
Lanius cristatus cristatus

Size: 19cm

Habitat: Scrubby habitats.

Distribution: Spreads throughout Sri Lanka on arrival.

Voice: Harsh rattle.

Status: Common Migrant.

Brown Shrikes, especially juveniles, can have a pale brown crown, edged pale on the sides and with a fairly broad white forehead. Viewed from certain angles, especially from the side, whole crown can appear to be silvery-grey, leading to confusion with Philippine Shrike (see right). Juveniles have barring on underparts and broad, pale-edged tertials. Brown Shrikes in different stages of age and moult can look very different. Sometimes they can look dull; at other times they show rich orangeish-buff on flanks and look richer brown above.

BROWN SHRIKE
(PHILIPPINE SHRIKE)
Lanius cristatus lucionensis

Size: 19cm

Habitat: Scrubby habitats.

Distribution: Spreads throughout Sri Lanka on arrival.

Voice: Harsh rattle.

Status: Scarce Migrant.

Grey on Philippine Shrike (a race of Brown Shrike, see left) is quite extensive from crown to mantle. Very good views of Philippine make it easier to be sure of race. Juveniles have barring on underparts.

BAY-BACKED SHRIKE
Lanius vittatus

Size: 17cm

Habitat: Frequents scrub jungle.

Distribution: First recorded from Bundala, and a bird was subsequently seen for a few consecutive years in Yala National Park. Records also from Uda Walawe National Park. Records from the South-east may be a reflection of these being heavily watched areas, as suitable habitat is also found in northern half of Sri Lanka.

Voice: Contact call a shrill *kreek-kreek*. Song a varied melody of squeaky and chattering notes.

Status: Vagrant.

Looks similar to Long-tailed Shrike (see right) frontally, but has maroon not grey mantle. Whitish rump. Lower mandible has pale base.

LONG-TAILED SHRIKE (RUFOUS-RUMPED SHRIKE)
Lanius schach caniceps

Size: 25cm

Habitat: Scrub jungle areas.

Distribution: Northern parts of Sri Lanka. In areas such as Mannar, not difficult to see on roadsides. It is unusual that it has not spread further south given the availability of suitable habitat.

Voice: Harsh, repeated *kraa*.

Status: Scarce Resident.

Looks similar to Bay-backed Shrike (see left) frontally, but has grey not maroon on mantle. Rufous rump. A fairly confiding shrike that allows a close approach and has become habituated to people.

SOUTHERN GREY SHRIKE
Lanius meridionalis

Size: 25cm

Habitat: Bird of arid, stony lands and sparsely vegetated open country.

Distribution: Most likely to be seen in dry lowlands.

Voice: Repeated, double-noted, hollow-sounding *treep-treep*. Also repeated *trow-eek* with first part low and second part high in pitch with a reedy quality. Song includes a repeated, nasal *zhi-wee*.

Status: Vagrant.

Also known as Iberian Grey Shrike. Overall impression is of a grey, black-and-white shrike. Grey on upperparts, black wings with large white primary patch and black tail with white outer feathers. Black mask separates grey crown from white face. Great Grey Shrike *L. excubitor* is similar, but Southern Grey is darker above, with flanks having pinkish-grey tinge. Southern Grey has been spilt from Great Grey, and some races of Great Grey have a small primary white wing-patch as in Southern Grey, while others have large primary patches. Southern Grey has distinctive curly white eyebrow. Lesser Grey Shrike *L. minor* (not recorded in Sri Lanka) is also similar, but separated by having a black forehead (grey in Southern Grey), and salmon-pink on breast and belly.

Thrushes Family *Turdidae*
Most thrush species are drab and forage in shaded undergrowth – but there are exceptions, which are brightly coloured. Some thrushes are among the finest of songsters, and they make hissing contact calls.

PIED GROUND-THRUSH
Zoothera wardii

Size: 22cm

Habitat: Forested areas in hills and highlands.

Distribution: Mid-hills to highlands. Victoria Park in Nuwara Eliya is a reliable site. Fond of berries of introduced Japanese Mahonia *Mahonia japonica*. Plants in *Mahonia* genus and in Berberidaceae family occur naturally in its breeding grounds in the Himalayas, and this may explain why it favours Japanese Mahonia's berries.

Voice: Song described as having high-pitched notes (I have not heard it sing in Sri Lanka).

Status: Uncommon Migrant.

Striking black-and-white thrush with conspicuous white supercilium. Female has same pattern as male, but with black being replaced by brown.

ORANGE-HEADED THRUSH
Zoothera citrina citrina

Size: 21cm

Habitat: Forest patches and occasionally well-wooded gardens with plenty of leaf litter.

Distribution: On arrival spreads from lowlands to mid-hills.

Voice: Rich, fluty notes.

Status: Scarce Migrant.

Orange head and upperparts. Grey wings, mantle and tail. Mantle, wings and tail of female are browner than male's.

Female

SPOT-WINGED GROUND-THRUSH
Zoothera spiloptera ⓔ

Size: 21cm

Habitat: Densely shaded forests. Forages for invertebrates on forest floor.

Distribution: Wet zone, lowlands to mid-hills. Emerges to feed on trails at dawn and dusk.

Voice: One of Sri Lanka's five species of true songbird, endowed with a double larynx. Wide repertoire of call notes and rich, melodious song.

Status: Endemic.

White spots on wings are easy to make out. Face marked strongly with black and white.

CEYLON SCALY THRUSH
Zoothera imbricata ℮

Size: <26cm

Habitat: Forests from lowlands to highlands. Partial to forest patches adjoining streams.

Distribution: Wet zone from lowlands to highlands where good-quality forest remains. Sinharaja is the best location for birders.

Voice: Hissing call more high pitched than that of Spot-winged Ground-thrush. Seldom sings. Song reminiscent of but different from that of Spot-winged. In March 2002, the bird's song was recorded in the evening in Horton Plains – until then it had been thought that it sang only in the mornings.

Status: Scarce Endemic. Endangered on IUCN Red List.

Underparts brown with dark scaling unlike that of Spot-winged Ground-thrush (see p. 220), which has white underparts with black spots. Wings do not have two white-spotted bars of Spot-winged. Longer, heavier bill and lacks white spots on wings of Spot-winged. Elusive bird, rarely permitting a clear view.

INDIAN BLACKBIRD
Turdus simillimus kinnisii

Size: <26cm

Habitat: Bird of highlands; has adapted to human presence and visits home gardens. When the *'Nillu' Strobilanthes* spp. are in bloom, Horton Plains National Park seems to have a pair every hundred metres. At other times quite difficult to see in park.

Distribution: From mid-hills to highlands in wet zone.

Voice: One of five songbirds resident in Sri Lanka. Beautiful and complex repertoire of rich, liquid notes. Contact calls are typical thrush-like hissing notes.

Status: Resident.

The only all-black thrush. Bill redder than that of Blackbird *T. merula* found in Europe. Race *kinnisii* found in Sri Lanka may in future be split from race found in India to be a 'good species'. Orangeish legs and reddish eye-ring. Female browner than male.

Eyebrowed Thrush

Turdus obscurus

Size: 23cm

Habitat: In breeding grounds inhabits Siberian Taiga.

Distribution: Records from highlands, especially Victoria Park in Nuwara Eliya – a magnet for wintering thrushes.

Voice: Complex song with melodic whistles and bubbly notes. Reminiscent of European Blackbirds *T. merula*. Alarm calls are typical thrush-like bursts of compressed notes.

Status: Vagrant.

Adult male has grey head, black lores and broad white supercilium. Thick white line runs below black lores, creating an unusual facial expression. First-winter bird has brown head, but there is a duller impression of eyebrow and pattern of loreal region. In adult and first-winter bird, breast and flanks are orange, and belly is white. Adult female like browner juvenile, but lacks juvenile's white tips to greater coverts. All birds have orange legs. It is not possible to sex birds with certainty, though females can be expected to be duller than males.

Ceylon Whistling-thrush

Myophonus blighi **e**

Female

Male

Size: 20cm

Habitat: Densely shaded streams. Mainly insectivorous, but opportunistically eats vertebrate animals such as geckos and agamid lizards.

Distribution: Confined to cloud forests in Central Highlands and Knuckles. Also recorded in Sinharaja East.

Voice: Vocalizes in early morning or evening, when its shrill, grating call, *sree sree, sree sree*, helps to locate it.

Status: Scarce Endemic.

Male brownish-black with blue gloss on foreparts and blue shoulder-patch. Female brown with blue shoulder-patch. In densely shaded undergrowth favoured by the birds, can look dull black. Whistling-thrushes are sensitive to ultraviolet light, and pattern of contrasting blue may be much more vivid to them than it appears to us.

Old World flycatchers and chats Family
Muscicapidae
A mixed bag of species comprises this family.
Flycatchers typically return to a hunting perch
in the canopy. Chats typically prefer the shaded
undergrowth, often foraging on leaf litter. Many
species make *tic tic* notes.

ASIAN BROWN FLYCATCHER
Muscicapa dauurica

Size: 14cm

Habitat: Bird of the canopy.
Occupies wooded gardens even in
cities, where it can be seen flitting in
search of insects.

Distribution: Spreads throughout
Sri Lanka from lowlands to mid-hills.

Voice: Single-noted *tchree* call.

Status: Migrant.

BROWN-BREASTED FLYCATCHER
Muscicapa muttui muttui

Size: 13cm

Habitat: Favours wet lowland
forests, but also frequents riverine
vegetation in dry zone.

Distribution: Spreads widely
from lowlands to mid-hills where
shaded forests are found. Absent in
highlands.

Voice: High-pitched *sree* contact call. Song very soft, as
if it is singing to itself.

Status: Migrant.

Dark legs are diagnostic in distinguishing this
species from Brown-breasted Flycatcher, which
has flesh-coloured legs. Habitat preferences are
also different. Both Asian Brown Flycatcher and
Brown-breasted Flycatcher (see right) have a
lower mandible that is pale and tipped dark.
Brown-breasted is darker above and has a browner
breast-band. Calls are also diagnostic.

Pale legs separate this species from Asian Brown
Flycatcher (see left), which has black legs. Darker
brown above and darker brown breast-band. Both
species have pale lower mandible.

YELLOW-RUMPED FLYCATCHER
Ficedula zanthopygia

Size: 13cm

Habitat: Inhabits undergrowth in vegetation lining streams and rivers on lower bases of valleys in East Asia.

Distribution: May occur in any river valley in mid-hills or lowlands.

Voice: Song variable and comprises a series of high strophes. Some notes are thrush-like.

Status: Vagrant.

Adult male is striking, with yellow on rump and underparts, black on tail and rest of body, white eyebrow and white patch on wings. Female and first-winter male have yellow rump, but lack any black and have greenish-grey upperparts and dirty-white underparts.

Male

KASHMIR FLYCATCHER
Ficedula subrubra

Size: 13cm

Habitat: Mainly in highlands, but at times in mid-hills. Occupies mid-storey of trees.

Distribution: A large part of the world's Kashmir Flycatchers winter in highlands of Sri Lanka. Victoria Park in Nuwara Eliya is a well-known site. A confiding bird.

Voice: Contact call a rattled note. Song a whistled *chip chip* followed by a rattling note.

Status: Scarce Migrant.

Slaty-grey upperparts. Male has black line running from bill along sides of head and bordering red throat and breast, which fades into dirty white of belly. Female lacks red on throat and breast. Both sexes have white at base of tail. Habit of raising and lowering tail.

DUSKY BLUE FLYCATCHER
Eumyias sordidus ℮

Size: 14cm

Habitat: Forest bird of highlands, found occasionally in mid-hills. Has adapted to human presence and also visits gardens.

Distribution: Largely confined to highlands. Also present in submontane areas of Eastern Sinharaja and where dry zone meets northern slopes of Knuckles Mountains.

Voice: Light, lilting song that appears to be coming from further afield than it really is. Notes reminiscent of a blackbird's song.

Status: Uncommon Endemic. Vulnerable on IUCN Red List.

Blue flycatcher with duller blue underparts fading to dirty white near vent. Loreal region is black. Vocalizations, behaviour and profile different from those of energetic Black-naped Blue Monarch (see p. 211), with which it is sometimes confused. Lacks black nape.

Female

Male

BLUE-AND-WHITE FLYCATCHER
Cyanoptila cyanomelana

Size: 16–17cm

Habitat: Forest patches where trees are present. Swoops to the ground to catch prey, then returns to perch.

Distribution: One vagrant record in Sinharaja forest in March 2014 was the second record on the Indian subcontinent of this bird, which breeds in East Asia (North and South Korea, and Japan) and winters in Southeast Asia. Vagrant records also from Oman, Middle East.

Voice: Song variable. Typically a mix of thin, musical *tee-tee-teew* notes, sometimes on an ascending scale.

Status: Vagrant.

Blue upperparts; bluish-black on face, chin and throat extends down to upper breast. Clean white on rest of underparts, including vent. Female has brown upperparts, pale chin and throat, and breast and flanks that diffuse into white belly and vent. In female, white is not clean-cut; in male it is sharply demarcated. Both sexes differ from resident blue flycatchers in Sri Lanka. Adult and first-winter males have white sides at base of tail that show in flight.

BLUE-THROATED FLYCATCHER
Cyornis rubeculoides

Size: 14cm

Habitat: In breeding range, found in well-wooded areas and forests in mountainous areas.

Distribution: Recorded in dry lowlands and hill country in Sri Lanka. May be overlooked for more common Tickell's Blue.

Voice: Includes ascending jingle of notes starting off with double-noted *whoo whoo* and ending in tremulous *tirrup*. Similar in tone and structure to voice of more common Tickell's Blue, but sufficiently distinct to be told apart. Other song sequences very different from those of Tickell's Blue. Passerine vagrants to Sri Lanka may not sing in their winter quarters.

Status: Vagrant.

Male

Male superficially similar to male Tickell's Blue Flycatcher (see right) and may be overlooked. As its name suggests, has a blue throat that has a well-defined demarcation with orange breast. In Tickell's Blue, orange extends up to throat and is bordered on face by black area that extends up to eye. Females are different. Female Tickell's Blue looks like a duller version of male, but female Blue-throated is brown on upperparts and has a white throat with a diffuse pale orange breast-band that shades into white belly. Female has pale eye-rings and paler lores on brown face.

Male

TICKELL'S BLUE FLYCATCHER
Cyornis tickelliae jerdoni

Size: 14cm

Habitat: Confiding bird of forest patches throughout lowlands ascending to mid-hills.

Distribution: Widespread from lowlands to mid-hills where shaded forests are found. Not uncommon in dry zone, but found only in tall, shaded forests, and absent from more open scrubland.

Voice: Tinkling series of five or more notes. End of a sequence marked by a *pip-pip*. A beautiful little melody.

Status: Resident.

Male has blue upperparts, and black on face bordering orange throat. Orange continues to lower breast and fades into white underparts. Female lacks black on face and is duller on underparts. Tickell's Blue may be confused with vagrant Blue-throated Flycatcher (see left). Males are similar, but latter has blue not orange throat. Female is brown not blue on upperparts, and has whitish lores.

GREY-HEADED CANARY-FLYCATCHER
Culicicapa ceylonensis ceylonensis

Size: 13cm

Habitat: Hills and highlands, forests and gardens.

Distribution: Main range is in highlands, where it is a common bird in cloud forests. In cities such as Nuwara Eliya it is a garden bird, but this may be because the city is a cloud-forest city, set within a bowl surrounded by cloud forests.

Voice: Machine gun-like chattering call that is similar to call of a canary, hence the name canary-flycatcher.

Status: Uncommon Resident.

Sparrow-sized bird with grey head and breast, olive-yellow above and brighter yellow below. Confiding bird that will perch in view and allow a close approach.

Male

BLUE ROCK-THRUSH Monticola solitarius

Size: 23cm

Habitat: Rocky habitats including mountains and steep cliffs near coast.

Distribution: Rocky areas in dry lowlands and potentially almost anywhere in Sri Lanka. I have found it in rocky outcrops in the Sithulpahuwa temple complex in the Yala protected area in the South-east. It is not shy of people, and on arrival in Sri Lanka can turn up in gardens.

Voice: Song variable and includes a thin, musical series of notes reminiscent of those of common Oriental Magpie-robin (see p. 229). Other songs include a series of short, sharp notes and tremulous notes at the end.

Status: Scarce Migrant.

Long-billed, long-tailed thrush that looks featureless. Adult male blue overall, but in most situations blue is not obvious and bird looks dark. Female brown overall, with underparts lighter and barred darker brown. Juvenile male blue overall with scaly appearance, with light edges to feathers and light barring underneath.

BLUETHROAT Luscinia svecica svecica

Size: 15cm

Habitat: In breeding range occupies swampland and riverbanks with dense vegetation. In wintering range looks for similar waterside habitats.

Distribution: May turn up anywhere with suitable riverine habitat.

Voice: Song highly variable. Song sequences include series with reedy *cheeps* mixed with *tchlips* and rattling notes.

Status: Vagrant.

Both sexes have white eyebrow, black eye-line and rufous edge to base of tail. Male in breeding plumage has blue chin and throat extending to upper breast, with a rufous patch in the middle. Adult females and first-winter males have less blue, with a white throat and white moustachial stripe. First-winter females have no blue.

Female

Male, winter

Male

Female

INDIAN BLUE ROBIN Luscinia brunnea

Size: 15cm

Habitat: Hills and highlands. Skulks in dense undergrowth.

Distribution: Spreads all over Sri Lanka except in the most arid areas and coastal regions. Number of birds increases with elevation. In hills, every wooded thicket seems to have one or more of these birds, but they are very difficult to see. Victoria Park in Nuwara Eliya is the top site for birders.

Voice: Series of high-pitched *hee i* notes that end with a set of 2–3 quivering notes. Once learnt, it becomes easy to gauge how many individuals of this difficult-to-see bird are present in winter.

Status: Migrant.

Male has blue upperparts, white supercilium and red underparts fading to white on vent. Facial area is blackish. Female has brown upperparts and underparts are a lighter version of male's.

Rufous-tailed Scrub-robin
Cercotrichas galactotes

Size: 17cm

Habitat: Bird of dry, open country. In its breeding range found in fruit orchards.

Distribution: First record was from urban Colombo, but preferred habitat is in dry lowlands.

Voice: Musical song with a variety of notes, some tremulous and some metallic. Typically robin-like in varied structure of song. Some song notes have screechy notes at the end.

Status: Vagrant.

Upperparts warm buff. Face has a thick, pale supercilium, dark eye-line, dark moustachial stripe and hint of a dark malar stripe. Rufous tail has white tips edged with black at bases. Black-and-white spots can be seen when tail is spread. Often jerks tail when on top of a perch. Uppertail feathers are muddy brown.

Male

Female

Oriental Magpie-robin
Copsychus saularis ceylonensis

Size: 20cm

Habitat: Feeds primarily on the ground, though it uses high perches for singing. More likely to be found in degraded habitats than in primary forest. Common garden bird.

Distribution: Widespread from lowlands to highlands.

Voice: Has a double larynx, making it a true songbird. Inclined to sing until late in the evening after light has faded. Has a number of harsh, scolding calls.

Status: Common Resident.

Black-and-white chat with longish tail that it raises often. Male glossy black on head, mantle and tail. Female has duller black parts, almost a blackish-grey. Young are brown and spotted on breast.

WHITE-RUMPED SHAMA
Copsychus malabaricus leggei

Size: 25cm

Habitat: Favours fairly thick lowland forests.

Distribution: Mainly in dry zone from lowlands to lower hills.

Voice: One of five Sri Lankan songbirds with a double larynx. Complex song of whistled notes, chattering scolding calls and mimicry vocalizations borrowed from other birds.

Status: Resident.

Distinctive bird with glossy blue-black upperparts, long, graduated tail, conspicuous white rump and orange-red underparts. Sexes similar, with female having slightly shorter tail. Does occasionally perch in the open, but is most often surprisingly difficult to see as it prefers to sing from within a tangled thicket, just like Nightingale *Luscinia megarhynchos* of Europe – another accomplished songster. The shama often sings till late, its song overlapping with nightjars churring at dusk.

Male

Female

INDIAN BLACK ROBIN
Saxicoloides fulicatus leucoptera

Size: 16cm

Habitat: Lowlands and hills, forests and gardens. Occasionally, a few birds even take up residence in Colombo.

Distribution: Widespread except in highlands. Most common in dry lowlands. Common bird in national parks of dry lowlands.

Voice: Repeated *chizwee* alternated with double note. Female responds to male in lower pitch.

Status: Resident.

Male glossy black with white shoulder-patch and reddish-brown vent. Female brownish-black with rusty vent.

WHINCHAT *Saxicola rubetra*

Size: 12cm

Habitat: Open grassland, usually near waterbodies.

Distribution: Typically a bird of low elevations, likely to be seen in grassland surrounding man-made lakes in dry lowlands. May turn up in wet zone as well.

Voice: Song high-pitched and scratchy with rapidly delivered notes of variable intonation. Some notes have a trill.

Status: Vagrant.

Male

Both sexes and juveniles show a prominent pale supercilium. In breeding male, supercilium is white and contrasts with blackish face. Females lack blackish face. Crown streaked in all plumages. Broad white base to tail with dark, wide terminal band. Breeding male has two white wing-patches.

Male

SIBERIAN STONECHAT
Saxicola maurus

Size: 13cm

Habitat: Occupies grassland and heathland in low-lying areas as well as in mountains.

Distribution: May turn up anywhere in grassland, from lowland sites such as Uda Walawe National Park to Horton Plains.

Voice: High-pitched, scratchy song with various *tchwee* notes delivered in different ways.

Status: Vagrant.

Recently split from Common Stonechat *S. torquatus*, the species occurring in Sri Lanka is believed to be Siberian Stonechat. In adult breeding male Siberian, underwing-coverts are jet-black, whereas in Common they are grey and underwing shows no contrast. In Siberian, white patch below hood of head is very large compared to Common's. Other characteristics, such as slightly longer wings, are not easily discernible in the field. Caspian race 'stejnegeri' has more white on tail base.

PIED BUSHCHAT
Saxicola caprata atrata

Size: 13cm

Habitat: Open areas interspersed with bushes and trees in highlands.

Distribution: Grassland bordered with forest in highlands.

Voice: Repeated soft *chiew* contact call. Soft song of a repeated tremulous triple note a characteristic sound of highland landscapes.

Status: Uncommon Resident. Endangered on IUCN Red List.

Male

Female

Male black and white with prominent white wing-bar, and white rump and vent. Female light brown with rusty rump. Female Indian Black Robin (see p. 230) lacks rusty rump, has a longer tail and behaves more like a magpie-robin. The two do not share the same range.

PIED WHEATEAR
Oenanthe pleschanka

Male, summer

Female

Size: 15cm

Habitat: In breeding range favours barren lands with boulders and short grass. Can be found on mountain slopes and high plateaux.

Distribution: Dry lowlands would be the most likely places to see it, though the first record in Sri Lanka was from a garden in Colombo.

Voice: Song a high-pitched and thin medley of short, sharp notes, some punctuated with short gaps. A few longer strophes.

Status: Vagrant.

Deep stem to inverted T-shape on tail. Breeding male has white cap extending from top of head to nape. Crown dirty on pale cap. Face, chin and throat, mantle, back and wings jet-black. Adult breeding female nondescript brown, pale on belly with a darker breast, head and upperparts. Breast can show streaks. Juveniles similar to adult female, but have contrasting pale edges to wing feathers and are scaled on back.

DESERT WHEATEAR *Oenanthe deserti*

Size: 15cm

Habitat: Arid lands with little vegetation. Avoids pure sandy deserts.

Distribution: Dry lowlands are the most likely place to see this vagrant.

Voice: Songs include a sequence that begins with *tee-tivoo* and ends in a trill. Some very high-pitched, rapid notes.

Status: Vagrant.

Black on tail extends almost from base to tip. An inverted T-shape is absent. Underparts pale. In breeding plumage, black wings and face contrast with warm buffish upperparts. Breeding male has white scapulars forming white panel against wings.

Male acquiring summer

Male acquiring summer

Winter

ISABELLINE WHEATEAR
Oenanthe isabellina

Size: 16cm

Habitat: Dry, boulder-strewn plains with short grass.

Distribution: Records have been from dry lowlands in northern half of Sri Lanka and in South-east.

Voice: Song a series of scratchy notes and half warbles. Quite variable. Some song sequences seem nervous and half finished.

Status: Vagrant.

Pale underparts and sandy upperparts. Black alula. The least contrastingly marked of the three wheatears recorded in Sri Lanka. Both sexes and first-winter birds have pale or white eyebrows. Breeding male has black lores. Short stem to inverted T-shape on tail. Pied Wheatear (see p. 232) has a deep stem and Desert Wheatear (see left) lacks inverted T-shape on tail. Characteristic erect posture.

> **Babblers** Family *Timaliidae*
> Most babblers live in tightly knit flocks and are highly vocal. In some species a dominant pair has helpers that assist with incubation and raising young. They occupy a variety of habitats, from primary rainforest to savannah.

ASHY-HEADED LAUGHINGTHRUSH
Garrulax cinereifrons (e)

Size: 23cm

Habitat: Confined to extensive lowland rainforests.

Distribution: The 'Barrier Gate' flock at Sinharaja offers a good chance to see the bird.

Voice: Flocks keep up a medley of 'hysterical' sounding calls, sometimes with a faint metallic quality.

Status: Scarce Endemic. Endangered on IUCN Red List.

Brown body and bluish rather than ashy head, which may not be noticed unless a good view is had. Typically forages near the ground and may be missed in a Sinharaja Bird Wave unless you are alert to its calls. Flocks are quite habituated in Sinharaja and permit good views, especially on the rarer occasions when they explore the mid-levels of the canopy or come out to feed on '*bowitiya*' bushes.

BROWN-CAPPED BABBLER
Pellorneum fuscocapillus (e)

Size: 16cm

Habitat: Where forest patches remain.

Distribution: Found throughout Sri Lanka up to highlands. Widespread bird whose call can be heard from every forested thicket.

Voice: Distinctive *pritee dear* betrays the presence of the bird, which is a great skulker, rarely showing itself in the open. Beautiful song.

Status: Uncommon Endemic.

Small brown babbler with darker brown cap. Very shy and keeps hidden. In Sinharaja, has become quite habituated and pairs may come out on to the road in the evenings.

CEYLON SCIMITAR-BABBLER
Pomatorhinus [schisticeps] melanurus e

Size: <22cm

Habitat: Favours areas of good forest. Found in village-garden habitats where areas of forest adjoin.

Distribution: Found in both lowland wet zone and dry zone, and ascends to highlands. However, in dry zone is absent from thorn-scrub forest.

Voice: Almost always found in a duetting pair: male utters a long, bubbling series of calls that end with a *kriek* from female. Song is so well synchronized that the sound often appears to come from a single bird.

Status: Endemic.

Prominent white eyebrow, black mask though eye and downcurved 'scimitar' bill help to identify this species. Dark brown upperparts contrast with white underparts.

TAWNY-BELLIED BABBLER
Dumetia hyperythra

Size: 13cm

Habitat: Grassland and scrub habitats.

Distribution: Widely distributed except in highlands, but common in grassy patanas of mid-hills.

Voice: Calls are explosive, high-pitched *tchee* or *zee* notes, followed by *kak-kak* and tremulous notes.

Status: Uncommon Resident.

Small babbler with brown upperparts. Underparts are tawny except for white throat. Nondescript, featureless babbler that is very restless and often seen threading its way through grassland on degraded hillsides.

DARK-FRONTED BABBLER
Rhopocichla atriceps

Size: 13cm

Habitat: Favours large forested areas, but even as at 2014, a small flock was holding out in the One Acre Reserve in Talangama on the outskirts of Colombo. However, this is atypical and the birds need to be close to forest patches.

Distribution: Throughout Sri Lanka.

Voice: Small flocks keep in touch with a *prruk prruk* chattering call.

Status: Resident.

Small babbler quite different in shape from larger babblers as it has a short tail. Highwayman's black mask, white eye-ring and pale bill lend it a distinctive look. A flock of Dark-fronted Babblers seems to follow every Sinharaja Bird Wave, but these birds hold territories and join a feeding flock as it sweeps through.

YELLOW-EYED BABBLER
Chrysomma sinense nasale

Size: 18cm

Habitat: Tall grassland and scrub.

Distribution: Found throughout Sri Lanka in suitable habitat up to hills, but absent from highest elevations. Most prevalent in dry lowlands.

Voice: Musical song quite unlike that of other species assigned to babbler family. Liquid-like fluty notes.

Status: Resident.

This looks like a 'misfit' in the babbler family, with its broad-based, short, stubby black bill. Orange-red eye-ring bordered with yellow, surrounded by face that is white except for brown cheeks. Brown upperparts, white underparts and yellow legs. Usually in pairs.

CEYLON RUFOUS BABBLER
Turdoides rufescens **e**

Size: 25cm

Habitat: Rainforests, mainly in lowlands but also in highlands. Tends to be found only where extensive forests remain.

Distribution: Wet zone. Almost absent from extensive but heavily disturbed Kanneliya Rainforest. Its near-absence is a mystery. In sites such as Kithulgala, forages in disturbed habitats, but always within a short flight from good-quality forest.

Voice: Chattering calls. Familiar sound in large lowland rainforests such as Sinharaja as it is a key nucleus species of the Sinharaja Bird Wave.

Status: Uncommon Endemic.

Orange bill and legs, rufous body and constant chattering help to distinguish this species from other babblers.

YELLOW-BILLED BABBLER
Turdoides affinis taprobanus

Size: 23cm

Habitat: Grassland and scrub in dry lowlands. Has adapted to home gardens.

Distribution: Widespread. There is no nook and corner in Sri Lanka that this babbler does not occupy, other than where high-rise buildings leave it without tree cover,

Voice: High-pitched, chattering calls. Uttered at a slow pace when foraging, they reach a high frequency and sound hysterical when birds are interacting.

Status: Common Resident.

Sandy-brown, almost grey babbler that is highly vocal and gregarious. Yellow bill, facial skin and legs on an otherwise fairly uniform colour scheme.

> **Cisticolas, prinias and tailorbirds** Family
> *Cisticolidae*
> This family comprises long tailed, warbler-like birds,
> many of which occupy grassland habitats. Their
> songs are far carrying and relatively loud for such
> small birds.

ZITTING CISTICOLA *Cisticola juncidis*

Size: 10cm

Habitat: Grassland and paddy fields.

Distribution: Widespread
throughout Sri Lanka.

Voice: *Zit zit* call makes it easy to
identify.

Status: Resident.

Warbler of grassland, heavily streaked on crown
and upperparts, with white-tipped tail that it
fans out in flight, hence the alternative name of
Fan-tailed Warbler. Utters a *zit zit* call in flight,
hence the onomatopoeic name. In short-cropped
grassland it can scurry about on the ground in a
mouse-like fashion.

GREY-BREASTED PRINIA
Prinia hodgsonii leggei

Size: 11cm

Habitat: Scrub forest in dry
lowlands. Most arboreal of the
prinias. Does not need grassland
like other prinias.

Distribution: Lowlands to mid-hills.
Easiest to see in dry lowlands.

Voice: Rattly three notes of *zeet
zeet zeet* uttered rapidly and repeatedly.

Status: Uncommon Resident.

Dark grey hood with clean white chin and throat.
Male has broad grey breast-band; female's is more
diffused. White underparts. Red irides gleam
through blackish loreal band, which continues
through eyes. Unlike other prinias, which are
usually seen alone or as a pair, Grey-breasted is
usually seen in small, noisy flocks.

ASHY PRINIA
Prinia socialis brevicauda

Size: 13cm

Habitat: Lowland grassland, but more common in patanas in hills. Can occur close to urban habitats. Next to Plain Prinia (see p. 240), the most resilient to urbanization.

Distribution: Found throughout Sri Lanka.

Voice: Clear, repeated *chewok chewok*. Sometimes higher pitched, repeated notes are introduced.

Status: Resident.

Bluish-grey upperparts; underparts tinged orange, paler on throat. Pale-tipped tail has black subterminal band. Female develops white supercilium in front of eye in breeding season. Reddish irides.

Female

Female

JUNGLE PRINIA
Prinia sylvatica valida

Size: 15cm

Habitat: Scrub jungle in dry lowlands.

Distribution: Found in suitable habitat throughout Sri Lanka except in highlands. Dry lowlands offer the best chance of seeing this bird.

Voice: Loud, repeated *tcheoow* note. Repeated note is strongly articulated, as if each note is thrown out with some force.

Status: Resident.

Diffused supercilium does not extend behind eye in male. Supercilium in both sexes less distinct in comparison to that of Plain Prinia (see p. 240). In the breeding season, female's supercilium extends behind eye. Stouter bill (dark in male) also distinguishes it from Plain Prinia. Reddish irides, black pupils. Calls of prinia species are different.

PLAIN PRINIA Prinia inornata insularis

Size: 13cm

Habitat: Lowland grassland.

Distribution: Widespread throughout Sri Lanka. The most tolerant of all prinias to urbanization. Can still be found close to Colombo in Kotte marshes and Talangama Wetland.

Voice: Range of cheerful vocalizations; typically a *tleep tleep* with higher pitched notes at end of a sequence.

Status: Resident.

Distinct, broad pale supercilium. Overall pale sandy-brown; much paler underneath, appearing almost white underneath at times. Black bill.

COMMON TAILORBIRD
Orthotomus sutorius

Size: 13cm

Habitat: Occupies disturbed habitats, hedgerows and gardens.

Distribution: Widespread throughout Sri Lanka; common garden bird.

Voice: Can include *yip yip* notes alternating with *pik pik*, then *tchewk tchewk* and variations, but generally a note is repeated a few times, before switching to another.

Status: Common Resident.

Green upperparts, white underparts, long tail and rusty crown. Fine, downcurved bill. Very different in appearance and behaviour from other members in family. Female lacks extended central feathers. Has jerky movements, and is very vocal. Named for nest it stitches, encased between two or more leaves it has sewn together.

> **Old World warblers** Family *Sylviidae*
> The warblers are quite varied in behaviour –
> *Phylloscopus* warblers are arboreal and glean
> insects in the canopy, while *Acrocephalus* warblers
> are birds of reed beds. All are small birds.

SRI LANKA BUSH-WARBLER
Elaphrornis palliseri ⓔ

Size: 16cm

Habitat: Thickets in highland forests.

Distribution: The '*Nillu*' undergrowth in the forests around Nuwara Eliya. Horton Plains is a reliable place to spot it.

Voice: Pairs often keep in touch using a series of nasal *tszip tszip* contact calls.

Status: Uncommon Endemic. Endangered on IUCN Red List.

Dark warbler with skulking habits. Keeps close to forest floor. Male has red irides; female's irides are buff. Paler, buffy throat.

Male

Female

LANCEOLATED WARBLER
Locustella lanceolata

Size: 12cm

Habitat: Swamps fringed with reed beds or rank vegetation.

Distribution: May occur anywhere in wetland habitats with extensive vegetation.

Voice: Song a rattling call that sounds like a grasshopper calling. Pitch is maintained evenly throughout long-drawn-out call. Contact calls include rapidly repeated, harsh chakking calls.

Status: Vagrant.

Heavily streaked on breast and flanks. Tertials dark with well-defined pale edges. Faint eyebrow. Can be streaked on undertail-coverts like Grasshopper Warbler (see p. 242). Uppertail-coverts streaked, unlike plain rusty rump of Grasshopper Warbler.

GRASSHOPPER WARBLER Locustella naevia

Size: 13cm

Habitat: Dense brakes of grass and reed vegetation usually beside water. In breeding range found in young conifer plantations with clear areas.

Distribution: May turn up anywhere with reeds and grass near water.

Voice: Song a long-drawn-out rattle with rapidly repeated notes, sounding like a grasshopper. Lanceolated has a similar song. Song remarkably consistent in pitch and amplitude, and sounds as though it is mechanically generated. Contact call is *pik-pik*.

Status: Vagrant.

Pale eyebrow. Streaked lightly on breast and flanks; streaks less heavy than in Lanceolated Warbler (see p. 241). Plain tail. Streaked undertail-coverts. Does not have dark tertials with pale edges seen in Lanceolated. Uppertail-coverts streaked, unlike plain rusty rump of Grasshopper Warbler.

RUSTY-RUMPED WARBLER
Locustella certhiola

Size: 13cm

Habitat: Aquatic vegetation. Reed beds or dense brakes beside waterbodies.

Distribution: Spreads throughout Sri Lanka on arrival. I have heard it in the Bellanwila-Attidiya marshes, as well as in the Elephant Nook wetland in Lake Gregory in Nuwara Eliya.

Voice: Several vocalizations, including typical warbler-like *chick*s. Song, a rolling *trrrt*, is likened to the sound of a fishing line running out of a rod.

Status: Scarce Migrant.

Plain, unmarked rusty rump. Prominent white eyebrow. All but central tail feathers have off-white tips. Black longitudinal lines on back are a feature also found in Lanceolated and Grasshopper Warblers (see p. 241 and left). In Grasshopper, black lines on back are the least distinct of the three species.

BLYTH'S REED-WARBLER
Acrocephalus dumetorum

Size: 14cm

Habitat: Occupies shrubby vegetation.

Distribution: Spreads throughout Sri Lanka.

Voice: Constant *chak chak* notes help to confirm its identity.

Status: Common Migrant.

INDIAN REED-WARBLER
Acrocephalus [stentoreus] brunnescens

Size: 19cm

Habitat: Lakes fringed with reed beds.

Distribution: Restricted to dry lowlands. Best chance of seeing it seems to be in freshwater reed-fringed wetlands on southern coastal fringe.

Voice: Call note is *chrek*. Song a series of grating notes, squeaks and shrieks. Highly variable mix of notes.

Status: Uncommon Resident.

Plain brown upperparts and fairly well-defined supercilium. Best located and identified by *chakking* call. Furtive bird, keeping well within shrubby thickets as it forages. Resident Indian Reed-warbler (see right) occupies wetland reed habitats and has different vocalizations. The latter is longer tailed and more likely to show itself.

Long billed and long tailed in comparison to migrant Blyth's Reed- warbler (see left). Easily recognized by chattering calls and occupancy of reed beds. Clear supercilium. In Blyth's, supercilium extends further back. Short primary projection.

SYKES'S WARBLER *Hippolais rama*

Size: 12cm

Habitat: Dry plains and arid areas. Bubbly song.

Distribution: Spreads throughout dry lowlands, but highest numbers of wintering birds are in Northern Peninsula down to Mannar.

Voice: Call note is *chek*. Bubbly song.

Status: Scarce Migrant.

BOOTED WARBLER *Hippolais caligata*

Size: 12cm

Habitat: Scrub and wooded areas, often near water.

Distribution: Arid zone in North-west and South-east, but may turn up anywhere in dry lowlands.

Voice: Contact calls are short *cheks* or agitated *trrrr*. Song a jaunty mix of unmusical notes reminiscent of a reed-warbler's song.

Status: Vagrant.

Very similar to Booted Warbler (see right), from which it has been split. Pale brown on upperparts, pale underparts and greyish-brown on flanks. Bill longer than in Booted. Supercilium behind eye faint or absent. Primary projection very short; about a third or less than that of the tertial length. Lower mandible pale with only a faint, dark tip. Sykes's does not twitch wings at the same time as tail, unlike Booted. Sykes's typically has a slightly longer tail (accentuated by short primary projection) and is slightly paler above. More horizontal stance like that of an *Acrocephalus* warbler.

Similar to Sykes's Warbler (see left), with grey-brown upperparts and paler underparts with rusty-buff tinge to flanks. Bill short, often with distinct dark tip. Bill is one of the best characteristics for distinguishing this species from Sykes's in the field. Legs brownish-pink with dark toes forming 'boot'. Booted twitches tail and wings at the same time.

Dusky Warbler *Phylloscopus fuscatus*

Size: 11cm

Habitat: Scrubby habitat interspersed with bushes and grass. Favours locations near water.

Distribution: Waterbodies in wet lowlands are the most likely locations.

Voice: Hard *chak* note. Song a repeated series of *chweek-chweek* or *tchlip-tchilp* notes. Sounds similar to Common Tailorbird (see p. 240).

Status: Vagrant.

Overall brown warbler. Distinct pale supercilium contrasts with brown crown. Distinct line through eye that is darker brown than that of general brown of upperparts. Chin and throat pale, shading into dusky underparts. Distinguished from Radde's Warbler *P. schwarzi* (not recorded in Sri Lanka) by bill being narrow and pointed (stout in Radde's) and plain ear-coverts (mottled in Radde's). Undertail-coverts uniformly coloured with flanks; in Radde's they are apricot-buff and are the brightest part of underparts.

Greenish Warbler
Phylloscopus trochiloides viridanus

Size: 11cm

Habitat: Frequents tree canopy in well-wooded forests In breeding range can occupy sub-alpine elevations.

Distribution: May turn up anywhere where there is suitable forest.

Voice: Contact call a repeated *chiwiz*. Mixed medley of short, sharp notes, including rapidly uttered *chee-chee* notes.

Status: Scarce Migrant.

Darker above than Bright-green Warbler (see p. 246), with greyish-white underparts lacking yellow underparts of Bright-green. Back greyish-green compared with moss-green of Bright-green. Typically only one pale wing-bar; both Bright-green Warbler and Large-billed Leaf-warbler (see p. 246) usually show a hint of a short second wing-bar. Short primary projection. Legs grey-brown. Habit of raising crown feathers. Supercilium broader behind eye and extends to forehead.

Bright-green Warbler
Phylloscopus nitidus

Size: 12cm

Habitat: Where adequate forest cover is present. Keeps mainly to canopy level on trees.

Distribution: Spreads throughout Sri Lanka. More abundant than Large-billed Leaf-warbler (see right).

Voice: Easily located by *thirrip* call.

Status: Common Migrant.

Similar to scarce Greenish Warbler (see p. 245), but has strong hint of yellow on underparts. Yellow wash reduces in worn adults.

Large-billed Leaf-warbler
Phylloscopus magnirostris

Size: 12.5cm

Habitat: Forests and well-wooded patches and gardens.

Distribution: Found throughout Sri Lanka. Appears to be absent from cities like Colombo, so may need forest canopy formed of native trees.

Voice: Calls are important in distinguishing leaf-warblers. Call structure is roughly of the form *hi-hee-hi*, which is repeated in variations of pitch. Call is thin. Song is musical and of the form *whi-we-weee-we-wee*, which is repeated.

Status: Common Migrant.

Dark eye-line and sharply defined supercilium against dark crown. Bill larger than Bright-green Warbler's (see left), but this is not very apparent in the field. Best way of distinguishing the two species is by their calls.

WESTERN CROWNED WARBLER
Phylloscopus occipitalis

Size: 11.5cm

Habitat: Found in forest canopy. In its breeding range found in mountainous areas.

Distribution: May potentially turn up anywhere with forests.

Voice: Call note a repeated *chitwee*. Song includes a series of high-pitched, short, alternating double-notes that are 'punched out'.

Status: Vagrant.

Dark green eye-line and prominent yellow eyebrow extending well behind eye. Crown olive with greyish-green central crown-stripe. Rump, tail and wings yellowish-green. Wings have two well-defined wing-bars. Mantle grey-green. Underparts greyish-white. Crown and mantle colour greyer than in Large-billed Leaf-warbler (see p. 246). Eastern Crowned Warbler *P. coronatus* (not recorded in Sri Lanka) has yellow undertail-coverts. Blyth's Leaf-warbler *P. reguloides* has broader wing-bars and dark panel on greater coverts.

INDIAN BROAD-TAILED GRASS-WARBLER
Schoenicola platyurus

Size: 18cm

Habitat: Hilly, open grassland.

Distribution: There are old sight records from mid-hills. A record of a specimen collected from 'Ceylon' is considered likely to be a mislabelling.

Voice: Contact call is a high, nasal *chik-chik-chik*. Song is in a similar vein, and includes repeated *chee-chee* notes.

Status: Vagrant.

Thin black lores with pale supercilium that extends to above ear-coverts. Irides brownish-grey. Upperparts dark ginger-brown. Underparts greyish-white. Some birds may be very pale on belly. Medium-length tail rounded at end. Legs and feet greyish-brown.

LESSER WHITETHROAT
Sylvia curruca halimodendri

Size: 12cm

Habitat: Farmland and wooded areas. Less likely to be seen in scrub or open country, unless there are thick hedges offering dense cover.

Distribution: Visitor to lowlands.

Voice: Song includes series of warbling notes ending with sequence of forceful and rapid piping notes. Less scratchy than Common Whitethroat's. Contact call a harsh *tchrrr*.

Status: Vagrant.

Lead-grey crown contrasting strongly with white throat, brown mantle and wings concolorous, clear white undertail-coverts and dirty flanks diffusing into off-white underparts. Easily separated from Common Whitethroat *S. communis* (not recorded in Sri Lanka), which has grey-brown mantle contrasting with rufous wings and paler grey on crown. Now split from Hume's Whitethroat (see p. 249), which is a larger and darker bird; mantle in Hume's is slaty-brown and crown is very dark, almost blackish. Dark grey legs and feet.

'DESERT' WHITETHROAT
Sylvia [curruca] minula

Size: <12cm

Habitat: Arid habitats with scrub.

Distribution: Arid zone in South-east. Highly likely to be recorded in North-east in Mannar area, through which migrants funnel into Sri Lanka.

Voice: Very similar to Lesser Whitethroat's. It may not be easy to separate calls and song in the field. Knowledge on this is still evolving, as 'Desert' Whitethroat is a split from Lesser Whitethroat.

Status: Vagrant.

Considered by some to be a race of Lesser Whitethroat (see left), and others as a 'good' species. Very similar to Lesser, but sandy grey-brown on upperparts. Crown lighter grey than in Lesser. Ear-coverts slightly darker. Looks like washed-out version of Lesser.

HUME'S WHITETHROAT
Sylvia althaea

Size: >12cm

Habitat: Scrubland interspersed with open woodland with bushes and short trees.

Distribution: Dry lowlands in more arid areas in South-east and North-east.

Voice: Song can be chirpy. Similar to songs of other whitethroat species, with much use of scratchy notes, cheeps and trills. Has a jaunty feel to it. Contact call a drawn-out *churrrr*.

Status: Scarce Migrant.

Separated from Lesser Whitethroat (see p. 248) by dark slaty back and much darker crown. Lores and ear-coverts may be darker than crown. Overall impression is of a darker bird on upperparts. Dark grey legs and feet.

Titmice Family *Paridae*
Titmice are small birds with a wide range of vocalizations. They are omnivorous, consuming invertebrates as well as seeds.

CINEREOUS TIT *Parus cinereous*

Size: 13cm

Habitat: Visits gardens in hills and highlands.

Distribution: Found throughout Sri Lanka. Common garden bird in highlands. In dry lowlands most likely to be seen in tall forests along rivers.

Voice: Contact call a rattling *churr*. Regularly used three-note contact call is *chee chee choo-wit*. Wide range of vocalizations, including far-carrying *teacher teacher* call.

Status: Resident.

Black head with bold white cheek-patch, and grey upperparts with prominent white wing-bar. Black chin and throat. Dark line runs through middle of belly to vent. Thicker and clearer in male than in female. Looks like grey form of familiar Great Tit *P. major* of Europe. Many authors treat this as a subspecies of Great Tit, which is widespread in Europe, and treat the bird found in Sri Lanka as Great Tit or Grey Tit *P. major mahrattarum*, distributed from West Asia, across South Asia to Southeast Asia. However, more recently there has been a view (adopted here) that it should be split as a separate species, Cinereous Tit *P. cinereous*.

> **Nuthatches** Family *Sittidae*
> Nuthatches are small birds with zygodactyl feet that can descend tree trunks head-first. Most are blue on the upperparts.

> **Flowerpeckers** Family *Dicaeidae*
> Flowerpeckers are stoutly built, small birds of the canopy. Their wide, deep bills help them to swallow berries that are relatively large in relation to their body size.

VELVET-FRONTED NUTHATCH
Sitta frontalis frontalis

Size: 10cm

Habitat: Tall and medium-stature forests, where it clambers along tree trunks, most times descending head-down. It can also be upside-down on a horizontal branch and still cling to the tree with its strong claws. Moves in a jerky fashion.

Distribution: Found throughout Sri Lanka.

Voice: Loud, quivering *chizzzz* call, made up of a series of machine gun-like *chizz* notes.

Status: Resident.

Female

Red bill, black forehead, dark blue upperparts and whitish underparts. Male has thin black line that continues behind eye; absent in female. Sometimes seen in family flocks, but typically adults are encountered in pairs.

THICK-BILLED FLOWERPECKER
Dicaeum agile zeylonicum

Size: 9cm

Habitat: Lightly wooded areas. Visits home gardens with trees in dry lowlands.

Distribution: North Central, Eastern and Uva Provinces, ascending central hills on drier side. Localized in wet zone.

Voice: Slightly tremulous, repeated *chlip chlip*.

Status: Uncommon Resident.

Reddish irides and moustachial stripe lend this species a very different look from Pale-billed Flowerpecker (see p. 252). Tail feathers are tipped with white.

LEGGE'S FLOWERPECKER
Dicaeum vincens  e

Size: 9cm

Habitat: Tall forests in lowlands. Habit of calling from tops of tall trees. Good views are had when it descends low to feed on fruits such as those from the common *Osbeckia* species along roadsides. Also takes invertebrate prey and nectar.

Distribution: Restricted to quality forests in wet lowlands. Visits village gardens adjoining good-quality forests, but it would be misleading to think that it has adapted to forest loss. Absent where forests have been heavily fragmented.

Voice: Males sing from high perch. Utters a plaintive, whistled *hi-hi-hi* ascending in tone, followed by a *pew-view pee-view*.

Status: Uncommon Endemic. Vulnerable on IUCN Red List.

White chin and throat, and yellow belly make it very different from the other two species of flowerpecker. Male has bluish upperparts, whereas female is duller with olive-grey. Male also has a stronger contrast with the white chin and yellow underparts.

Pale-billed Flowerpecker and Common Jezebel

There is an unusual connection between the Pale-billed Flowerpecker and the distribution of the beautiful Common Jezebel butterfly. Pale-billed Flowerpeckers are partial to the fruits of Loranthus plants. Sticky seeds evolved by the plants have provided them with a dispersal mechanism. In exchange for payment for high-energy food, the birds act as distribution agents for the offspring of the plants. The flowerpeckers visit home gardens and rub off the sticky seeds from their bills and bodies. In mango trees in particular, Loranthus plants take hold in the canopy. Loranthus is the food plant of the Common Jezebel, a gorgeous butterfly that has spread in cities like Colombo due to people planting mango trees and the flowerpecker spreading the plant. This is an example of how three species are explicitly linked in a chain of food and reproduction.

Male

Female

PALE-BILLED FLOWERPECKER
Dicaeum erythrorhynchos ceylonense

Size: 8cm

Habitat: Gardens and forests.

Distribution: Throughout Sri Lanka.

Voice: Contact call is *thlip thlip*. Song is a repeated, trilling *threep*. I have been awoken at dawn many a time by a bird singing from the top of a TV aerial being used as a song perch.

Status: Common Resident.

Overall impression is of a small grey bird, paler below. The small bill is distinctively downcurved. The tiniest bird on the island.

Sunbirds Family *Nectariniidae*
These are brightly plumaged, small birds, with most species having pronounced differences between the sexes. They feed primarily on nectar, and bill length can be very long in some species.

PURPLE-RUMPED SUNBIRD
Leptocoma zeylonica zeylonica

Male

Female

Size: 10cm

Habitat: Gardens and forest edges with nectar-rich plants. Mainly insectivorous, but feeds opportunistically on nectar from introduced garden plants, as well as native nectaring plants.

Distribution: Found throughout Sri Lanka up to mid-hills.

Voice: Contact call is a metallic *zeut*. Song is structured around a core *tseep* note with various notes alternating with it.

Status: Common Resident.

One of Sri Lanka's smallest birds, male is absolutely stunning with an iridescent green crown, yellow belly and glistening purple rump. Female is duller with brownish-grey head and dark line through eye. Both sexes constantly flick their wings.

Telling apart Purple and Loten's Sunbirds

	Loten's	Purple
Bill	Long	Shorter
Call	Metallic	More metallic than Loten's
Female	Thin supercilium	No supercilium
Eclipse adult male	No eclipse plumage	Dark line in middle of belly
Juvenile male	Black line extends from chin and throat to upper breast	Similar to female; no black line
Adult male underparts	Chocolate-brown or black	Glossy blue

PURPLE SUNBIRD
Cinnyris asiaticus asiatica

Size: 10cm

Habitat: Frequents home gardens, but also seen in thorn-scrub forests.

Distribution: Mainly in dry lowlands. Localized in wet zone.

Voice: Core note is a metallic *cheep cheep*, rising in intonation, followed by a rapidly repeated series of similar notes that end the call in a bubbling sequence. Notes are more metallic than those of Loten's.

Status: Resident.

Female

Male

Bill shorter than in Loten's Sunbird (see p. 254), and different calls. Male and female's colour patterns similar to those of Loten's. See table for additional features.

LOTEN'S SUNBIRD
Cinnyris lotenius

Size: 13cm

Habitat: Frequent visitor to gardens even in cities. Needs habitats where nectaring plants are easily available, and mix of nectar and invertebrates. Young are fed a rich variety of invertebrates, including spiders, caterpillars and similar animals.

Distribution: Most common in wet zone.

Voice: Series of loud, metallic notes, vaguely transcribing as *chewp chewp*, is used to announce its presence. In flight utters a rapid *chik chik chik*. Sometimes utters single-noted *cheep cheep* calls.

Status: Resident.

Male

Female

Breeding male glossy blue-black, at times iridescent green on head and mantle. Distinguished from similar male Purple Sunbird (see p. 253) by chocolate-brown underparts. In Purple they are glossy black. Female brownish-grey above and with yellow-tinged, pale underparts. Long, deeply curved bill. Bill shorter in Purple.

White-eyes Family *Zosteropidae*
White-eyes are small birds that are usually a shade of green or yellow on the upperparts, and white or pale below. The white eye-ring is a characteristic of the family.

CEYLON WHITE-EYE
Zosterops ceylonensis **e**

Size: 11cm

Habitat: Forests and wooded gardens.

Distribution: Mainly in highlands, with occasional movements to lowland wet zone (for example Sinharaja), where it mixes with Oriental.

Voice: The two white-eyes have different calls. Ceylon's calls are much louder than those of Oriental. Flocks utter chirping calls to stay in contact. Birds also have a call of a rapid sequence of *tchew tchew* notes that are drummed out.

Status: Endemic.

Has a wider 'split' in white eye-ring, in front of eye, than in Oriental White-eye (see p. 255). Also darker and slightly larger than Oriental. Calls of the two species are very different.

ORIENTAL WHITE-EYE
Zosterops palpebrosus egregia

Size: 10cm

Habitat: Occupies mid-storey and canopy of small trees.

Distribution: Islandwide. Most common in lowlands and hills, but can ascend to highlands.

Voice: Song a sequence of 6–7 complex notes. Birds keep in contact with series of soft, chattering calls.

Status: Resident.

Compared with Ceylon White-eye (see p. 254) has brighter lemon-yellow upperparts, cleaner white underparts and slimmer build. Calls are diagnostic in the field.

Buntings Family *Emberizidae*
These are small, sparrow-like birds with thick bills. Buntings in genus *Emberiza* are found across Asia, Africa and Europe. The family comprises more than 700 species in over 70 genera, and includes many of the 'American Sparrows' found across North and South America.

Male

GREY-NECKED BUNTING
Emberiza buchanani

Size: 15cm

Habitat: Bird of barren rocky hillsides; also visits stubble fields.

Distribution: In Sri Lanka, records are from dry lowlands, where open country and grassland are interspersed with scrub forest.

Voice: Song a repeated sequence beginning with *tsee-tsee-tsee*, followed by two falling *churoo-churoo* notes. Nasal, typical bunting quality. Contact call is a *chip*.

Status: Vagrant.

Grey head and yellowish throat pierced by grey lateral throat-stripe present in breeding male, female and first-winter birds. Bill pinkish-orange. Grey rump. Pale eye-ring. Similar to Ortolan Bunting E. *hortulana* (not recorded in Sri Lanka). Ortolan has grey breast-band. Female duller and buffish on head.

BLACK-HEADED BUNTING
Emberiza melanocephala

Size: 18cm

Habitat: Open country with bushes; flocks to ripening agricultural cereal fields and can be a pest in breeding range.

Distribution: In Sri Lanka, records are from dry lowlands where open country and grassland are interspersed with scrub forest.

Voice: Chirpy song is a repeated sequence ending with a clear, hard trilling *trrr*. Contact call is a *chlip*.

Status: Vagrant.

In breeding plumage male has black head, yellow underparts, and chestnut back and mantle spilling around wing shoulder. Chestnut rump. Heavy grey bill. Wings have black feathers with white edges. Non-breeding male duller than breeding male. First-winter birds are like females with streaking on mantle, breast and flanks.

Male breeding

Female

Male, non-breeding

RED-HEADED BUNTING
Emberiza bruniceps

Size: 17cm

Habitat: Scrub and arid open country. Visits cultivated areas in large flocks in breeding range.

Distribution: In Sri Lanka, records are from dry lowlands where open country and grassland are interspersed with scrub forest.

Voice: Song a repeated sequence starting with some hesitant *tszee-tszee* notes, and ending with a musical *chirp*. End of song sequence similar to that of Chaffinch *Fringilla colebs*, familiar to many visiting European birders (but not recorded in Sri Lanka). Contact call is a nasal *treep*.

Status: Vagrant.

In breeding plumage, male has red head and bib extending to breast, yellow underparts, and olive back and mantle. Rump is yellow and uppertail-coverts are olive. Wings have black feathers with white edges. Heavy grey bill. Non-breeding male is duller than breeding male. First-winter birds are like females with streaking on mantle, breast and flanks. Females and first-winter birds of Red-headed Bunting and Black-headed Bunting (see left) are similar – distinguished by olive-brown mantle in Red-headed and light chestnut mantle in Black-headed.

Finches Family *Fringillidae*
This family has a number of species across the Americas to Europe, Africa, the Middle East and Asia. No resident members of the family occur in Sri Lanka. Finches are small birds with stout, conical bills for eating seeds.

Waxbills Family *Estrildidae*
Waxbills (a.k.a. munias) are small birds with strong, deep bills suited to eating seeds. Their plumage can vary enormously from bright patterning to drab. They occur in flocks.

Male

COMMON ROSEFINCH
Carpodacus erythrinus roseatus

Size: 15cm

Habitat: Found in montane woodland in its home range, where it feeds on seeds and pollen.

Distribution: Recorded in Horton Plains National Park. Can winter in dry areas as well, so could potentially turn up anywhere in woodland.

Voice: Song clear and fluty – a collection of a few notes that at times can be like *chi-chu-wee-you-tchew*. Last note delivered with emphasis. Variable, and some song sequences are shorter than others.

Status: Vagrant.

Adult male has red head, and red extending from throat to breast and diffusing into white underparts. Red rump contrasts with greyish mantle, which is mixed with pink. Grey bill. Wings greyish with wing-coverts edged in pink. Female and juvenile males brown-grey with pale-edged wing-coverts. Juveniles have two buff wing-bars.

INDIAN SILVERBILL
Euodice malabarica

Size: 10cm

Habitat: Scrub jungle interspersed with open areas and grassland in dry lowlands.

Distribution: Confined to dry lowlands. Uncommon. Uda Walawe National Park is one of the best locations for it.

Voice: Strident chirps, reminiscent of House Sparrow (see p. 260).

Status: Uncommon Resident.

Overall impression is of a long-tailed, pale munia, looking pale grey or white overall. Underparts white with upperparts browner. Silver bill. Black tail contrasts with white rump.

WHITE-RUMPED MUNIA
Lonchura striata striata

Size: 10cm

Habitat: Lowlands and hills. Frequents clearings and degraded forest habitats.

Distribution: Lowlands to mid-hills.

Voice: Usual call is a twitter, sometimes preceded by a wheezy whistle. Occasionally intersperses twitters with whistled note.

Status: Resident.

Upperparts and head dark chocolate-brown, looking black at times. Mantle feathers have thin streaks. Underparts and rump are white. Bill has bluish hue.

BLACK-THROATED MUNIA
Lonchura kelaarti

Juvenile

Size: 10cm

Habitat: Mainly in highland forests with open areas. Occasionally in lower hills.

Distribution: Mainly in highlands, but occasionally seen in wet lowlands such as in Sinharaja. There may be seasonal movements in search of food.

Voice: Nasal, repeated *cheep cheep*. Quite distinct from calls of other munias.

Status: Uncommon Resident. Vulnerable on IUCN Red List.

Black throat and face, and scaly underneath like Scaly-breasted Munia (see p. 259), but much darker on upperparts. Can visit degraded sites such as Elephant Nook wetland in Nuwara Eliya, and confusion is possible with more common Scaly-breasted.

SCALY-BREASTED MUNIA
Lonchura punctulata punctulata

Size: 10cm

Habitat: Degraded forest areas in lowland, grassland and disturbed habitats that have tall grass colonizing them.

Distribution: Found throughout Sri Lanka. Occurs in large numbers in dry lowlands and considered a pest by farmers.

Voice: Distinct, clear *pit-teuw*, more of whistled tone than nasal chirp characteristic of most munias.

Status: Resident.

Brown head and upperparts; brown on throat up to breast. Underparts white with neat black scales. Vocalizes a lot, which helps to avoid confusion with rarer Black-throated Munia (see p. 258) where their ranges overlap.

TRICOLOURED MUNIA
Lonchura malacca

Size: 10cm

Habitat: Grassland.

Distribution: Found throughout Sri Lanka. Birds probably travel a lot in search for food, as small flocks can turn up even in small plots of wasteland in cities.

Voice: Series of repeated, nasal chirps.

Status: Resident.

Black head, chestnut upperparts, white breast and bordered around vent with black. Heavy, silvery bill. Juvenile plainer than Scaly-breasted Munia (see left) and also told apart by heavy, silvery bill.

> **Old World sparrows** Family *Passeridae*
> This is the family of the familiar House Sparrow. It
> comprises small birds with thick bills for eating seeds.

HOUSE SPARROW
Passer domesticus indicus

Size: 15cm

Habitat: Village garden areas that are edge habitats, with access to grain.

Distribution: Spread throughout Sri Lanka. Numbers have declined; a phenomenon observed worldwide. When I grew up in Colombo, rice was sieved to separate out chaff and stones in a flat wicker tray. The chaff and stones were cast aside on the ground. Earthen pots were hung as nest boxes for House Sparrows, so they had both food and nest sites. The change to use of supermarket ready-cleaned grain may be one of many factors accounting for the demise of House Sparrows in cities.

Voice: Series of repeated chirps. Also a *cheep-cheep-churp-churp.*

Status: Resident.

Male

Female

Male has grey crown and is more richly coloured than female. Fresh feathers on male wear off to reveal black bib during breeding season.

Male

YELLOW-THROATED SPARROW
Petronia xanthocollis

Size: 14cm

Habitat: Woodland, gardens and towns in breeding range.

Distribution: In Sri Lanka, a flock has been recorded in arid zone. May turn up anywhere in lowlands.

Voice: Song evenly spaced and chirruping, reminiscent of that of House Sparrow (see left), but faster and more rhythmic.

Status: Vagrant.

In both male and female, wing-coverts are bordered with 'black beads' and white wing-bar. In male, wing-coverts are chestnut and in female they are grey-brown, concolorous with head and rest of upper body. No streaking on body (crown, mantle and uppertail-coverts). Primaries dark in both sexes. Throat white in both sexes, but male has yellow patch below on lower throat (almost on upper breast). Both sexes have greyish breast-band. Stout bill is long (compared with short, stubby bill of most sparrows) and pointed. Bill black in breeding plumage.

> **Weavers** Family *Ploceidae*
> Weavers are seed-eating, sparrow-like birds. This family has evolved distinctive behaviour in that males build elaborate nests that are used to entice females.

STREAKED WEAVER
Ploceus manyar

Male   Non-breeding Male

Size: 15cm

Habitat: Reed beds, wetland or scrub near water.

Distribution: Small colonies found throughout dry lowlands.

Voice: Song a wheezy *whee-whee*, with mix of squeaky notes and at times ending with drawn-out note with slight trill.

Status: Uncommon Resident.

Both male and female can be distinguished from similar Baya Weaver (see right) by streaking on underparts. In breeding plumage, male acquires a yellow crown and is dark brown on face and throat. Non-breeding male and female similar to female Baya, but have streaks and conspicuous pale (at times yellowish) supercilium.

Male

BAYA WEAVER
Ploceus philippinus philippinus

Size: 15cm

Habitat: Thorn scrub bordering open clearings. At colonies, their distinctive bulbous nests with an entrance tunnel can be seen.

Distribution: Found throughout lowlands, but seems most prevalent in dry lowlands.

Voice: Chittering call interspersed with wheezing notes and strong chirps.

Status: Resident.

This species can be distinguished from Streaked Weaver (see left) by lack of strong streaking on breast in both sexes. Both sexes have pale edges to tertials. Female of Streaked also has white on supercilium. Female can be distinguished from female House Sparrow (see p. 260) by larger and stouter-looking, conical bill. Female weavers are darker above with heavy streaking, and lack white wing-bar found in female House Sparrow.

> **Starlings and mynas** Family *Sturnidae*
> Most starlings found in Sri Lanka are omnivores, eating a range of plant and animal matter. They happily take small berries, and spend a fair amount of time foraging on the ground for small invertebrates such as insects, earthworms and similar animals.

WHITE-FACED STARLING
Sturnornis albofrontatus ⓔ

Size: 22cm

Habitat: Frequents canopies of rainforests. Rarely seen in mid-canopy unless it has joined a mixed feeding flock.

Distribution: Restricted to a few lowland wet zone forests such as Sinharaja and Kithulgala. Starlings are powerful fliers, so may turn up briefly in small pockets of forest in search of food. However, it seems that breeding populations are confined to only large tracts of rainforest in lowlands and mid-hills.

Voice: Whistled *cheep* is repeated, sometimes turning into two-toned *cheep-cheowp*.

Status: Highly Scarce Endemic. Endangered on IUCN Red List.

GREY-HEADED STARLING
Sturnia malabarica

Size: 21cm

Habitat: Wooded areas and home gardens, where it frequents the tree canopy.

Distribution: May turn up in wooded areas in lowlands.

Voice: Calls a harsh and tremulous *krik-krik*.

Status: Vagrant.

White on forehead, face, supercilium, chin and throat. White on throat fades to dirty-white underparts with white streaks. Upperparts steely-grey. Overall impression when seen from below is of a pale bird.

Adults have grey upperparts, and chestnut underparts and tail. Throat white, contrasting paler than grey head. Juvenile grey-brown on upperparts, and paler underneath with chestnut sides to tail. Brahminy Starling (see p. 263) has dark cap and lacks chestnut tail.

PURPLE-BACKED (DAURIAN) STARLING *Sturnia sturnina*

Size: 19cm

Habitat: Lowland secondary forests, edges of primary forest and cultivated areas.

Distribution: Most likely to occur in lowlands.

Voice: High-pitched, tremulous *kee-kee-kee*. Also rapid *klip-klip-klip* and other slightly metallic calls.

Status: Vagrant.

Adult has purple patch on back and purple-sheen patch on nape. Broad white edge to scapulars and median coverts, and thin white edge on greater coverts. Secondaries largely black, primaries tipped black. White rump and underparts. Juvenile brown on back and scapulars. Lacks orange base to bill seen in juvenile Rosy Starlings (see p. 264). Juvenile has white undertail-coverts. Juvenile Rosy has pale-edged black undertail-coverts.

BRAHMINY STARLING
Temenuchus pagodarum

Size: 22cm

Habitat: Scrub jungle in dry lowlands

Distribution: Distribution is unusual for this winter migrant, as it favours arid zone in South-east and North-west, and generally coastal strip around northern part of Sri Lanka. Given that the thorn-scrub habitat is more widely spread, it is not clear why this passerine, which arrives in large numbers, occupies such a small area of dry lowlands.

Voice: Pleasant song; cackling contact calls.

Status: Migrant. Numbers variable from year to year.

Colourful starling with black crown and nape, orange-yellow bill with blue base, greyish upperparts and rufous-orange underparts. Pale irides are faintly green.

ROSY STARLING *Sturnus roseus*

Size: 23cm

Habitat: Scrub jungle in dry lowlands

Distribution: Can spread all over Sri Lanka in some years, but mainly in dry lowlands.

Voice: Song a series of chitters and chirps. Contact calls are quivering chirps.

Status: Migrant. In some years, tens of thousands of birds arrive and form huge roosts

Almost two-toned, with black hood extending down to upper breast, rosy underparts and mantle, and black wings. Orange-yellow bill. Should not be confused with Brahminy Starling (see p. 263), which does not have black hood and black on wings. Juveniles and first-winter birds (seen in Sri Lanka) are pale brown, and show traces of black plumage in second-year birds. Base of lower bill orange. Colours more vivid in breeding plumage.

COMMON MYNA
Acridotheres tristis melanosternus

Size: 23cm

Habitat: Garden bird that is a ground feeder on short-cropped grass. Turns over cow dung in search of insects and pulls out invertebrates in soft ground.

Distribution: Found throughout Sri Lanka.

Voice: Cheeps and chirps interspersed with grunting notes.

Status: Common Resident.

Yellow bill, yellow facial skin behind eye, and chocolate-brown upperparts and underparts, turning blackish on head. White vent and white wing-bar that show at rest. In flight wing-tips look a whirr of black and white.

CEYLON HILL-MYNA
Gracula ptilogenys ℮

Size: 25cm

Habitat: Frequents canopy.

Distribution: Bird of large, good-quality wet zone forests from lowlands to higher hills. Sinharaja and Kithulgala are good locations for this uncommon endemic.

Voice: Series of sharp-whistled *yowp* and *yeep* calls interspersed with nasal wheezing notes.

Status: Uncommon Endemic. Vulnerable on IUCN Red List.

Chunkier looking bird than Lesser Hill-myna (see right). Key differences are a single pair of wattles on nape (Lesser has extra pair under eye), black-based orange bill (yellow in Lesser), and pale, greyish irides (brown in Lesser). The way rear lappets are attached also differs. In Lesser there are two 'fingers' attaching lappet. Both species are overall glossy black birds with a prominent white wing-bar.

LESSER HILL-MYNA
Gracula indica

Size: 25cm

Habitat: Forests and well-wooded patches. Visits village gardens where there are good-quality forest patches nearby.

Distribution: Lowlands and hills in both wet and dry zones.

Voice: Long, whistled notes, separated by distinct interval with notes alternating in pitch, with a rising intonation followed by another with a falling intonation, interspersed with nasal sawing notes, *zee-zit*. Calls sound more complex than those of Ceylon.

Status: Resident.

Lesser Hill-myna has all-yellow bill (bill of Ceylon Hill-myna see left, is orange with black base). Ceylon is darker than Lesser. A diagnostic feature in Lesser is two pairs of wattles, one under eye (missing in Ceylon) and one on nape. In juvenile Lesser, nape-wattles develop after the one under eye.

> **Orioles** Family *Oriolidae*
> These are typically yellow birds marked with black, with liquid notes in their calls. They are arboreal in habit, and have adapted to town gardens.

EUROPEAN GOLDEN ORIOLE
Oriolus oriolus

Size: 25cm

Habitat: In breeding range found in deciduous woodland, or agricultural areas with copses and close to water.

Distribution: May occur anywhere in broadleaved woodland.

Voice: Harsh contact calls, including a slow-drawn *kreesh*. Also beautiful, melodic, fluty notes.

Status: Scarce Migrant.

Male

Golden body, largely black wings and tail with a lot of black. Key features for distinguishing this species from Indian Golden Oriole (see right) include: outer tail feathers have black bases (all yellow in Indian); in male, black loreal line does not extend behind eye (Indian male has black mask extending behind eye), and wing almost entirely black with a little yellow on tips of primary coverts. In Indian, secondaries and primaries have yellow tips and yellow patch on primaries. Orange-red bill. Female duller than male with faint streaking underneath. Juvenile more heavily streaked than adults, and yellow is replaced with dull green.

Male

Female

INDIAN GOLDEN ORIOLE
Oriolus kundoo

Size: 25cm

Habitat: In breeding range found in woodland, secondary forests and home gardens.

Distribution: In Sri Lanka occurs mainly in lowlands.

Voice: Contact calls are harsh screeches. Melodic, fluty double or triple notes.

Status: Scarce Migrant.

Similar to European Golden Oriole (see left), but differentiated by all-yellow outer tail feathers, black mask extending behind eye and much more yellow on wings, not just confined to small patch on primary coverts (see also description for European). Female duller than male. Juvenile heavily streaked and olive-green on upperparts.

BLACK-NAPED ORIOLE
Oriolus chinensis diffusus

Size: 25cm

Habitat: Secondary forests, plantations and home gardens.

Distribution: Recent records are from wet lowlands, surprisingly in urban areas.

Voice: Call a belling shriek. Song comprises fluty whistles.

Status: Vagrant.

Red bill, red irides, gold on body, wings mainly black and outer tail feathers basally black. Prominent black stripe from eye to nape. Slender-billed Oriole O. tenuirostris (no accepted records in Sri Lanka) is similar, but has greenish mantle and rump, and slightly curved, more slender bill.

Juvenile

BLACK-HOODED ORIOLE
Oriolus xanthornus ceylonensis

Size: 25cm

Habitat: Bird of the woods. Occupies well-wooded gardens even in cities. Omnivorous, feeding on both invertebrates (mainly insects, caterpillars and similar gleaned from trees) and fruits.

Distribution: From lowlands to mid-hills; absent in highlands.

Voice: Orioles have a rich vocabulary of calls. Some are harsh and grating, while others are very musical and fluty.

Status: Resident.

Adults have 'clean' black heads. Body yellow, wings and tail mainly black. Yellow edges to tertials and outer tail. Juvenile has streaky black head and duller colours. Vagrant Black-naped and Golden Orioles (see left and p. 266) lack black hood. In European Golden, black loreal line does not extend beyond eye. In Black-naped, thick black mask extends from base of bill to nape.

> **Fairy-bluebirds** Family *Irenidae*
> These medium-sized, strikingly coloured birds are
> sexually dimorphic. There are only two species in
> this family: Asian Fairy-bluebird, which is found
> from South Asia to Java and Palawan; and Philipine
> Fairy-bluebird *Irena cyanogastra*, restricted to the
> Philippines.

ASIAN FAIRY-BLUEBIRD
Irena puella

Size: 27cm

Habitat: Tall forests where it
searches for fruits. Also eats insects
and nectar.

Distribution: Most likely to occur in
wet zone, up to mid-hills.

Voice: Liquid notes, *weet-weet,
wick-we-wick.*

Status: Vagrant.

Male very striking with iridescent light blue on
crown, mantle and base of tail; iridescent black
on body with black wings. Female slaty-blue with
darker wings and tail. Both sexes have red irides.

Male

Female

Drongos Family *Dicruridae*
Drongos are typically all-black, insectivorous birds. Some species are prone to kleptoparasitism and use their powers of mimicry to alarm other birds with predator calls, making them drop their prey.

Telling apart the Drongos

	Black	White-bellied	Ceylon Crested	Racket-tailed	Ashy
Status and habitat	Resident in North-west, a few recent records in arid zone in South, and common in arid habitats close to coast	Resident, common bird; well-wooded gardens and degraded habitats	Resident in good-quality large tracts of forest in wet zone lowlands to mid-hills	Resident; prefers riverine forests in dry zone	Uncommon migrant to thorn-scrub forests in dry lowlands, and can also turn up in wet zone
Irides	Dull red	Dull red	Dark	Dark	Red
Forehead crest	NA	NA	Pronounced	Pronounced	NA
Racket tails	NA	NA	Forked tail ends with 'blades'; at times the feather sides may be abraded away, exposing the bare shaft and causing confusion with Racket-tailed Drongo	Pronounced rackets on long tail shafts	NA
Underparts	Black, vent white with black bars	Black except on vent and breast; dry zone race has more white extending from vent to belly, and wet zone race has white restricted to vent.	Black	Black	Black

BLACK DRONGO
Dicrurus macrocercus minor

Size: 31cm

Habitat: Thorn-scrub pockets in short-cropped grassland.

Distribution: North-western quadrant with occasional records from south-eastern Sri Lanka. Distribution is curious, and almost as if it is a recent colonist from Southern India that is spreading southwards from the Adam's Bridge connection to India. Common bird in Deccan Plateau of Southern India.

Voice: Chittering call that is slightly musical. Varied song with harsh *kraa* notes and whistles.

Status: Uncommon Resident.

White rictal spot near base of bill, but not always apparent. Irides duller red than those of Ashy Drongo (see right), which has crimson irides. Flight feathers brown. Under-tail coverts white barred with black, but do not always show.

ASHY DRONGO
Dicrurus leucophaeus

Size: 30cm

Habitat: Found in broadleaved forests as well as in conifers in breeding range. Like many drongos, often occurs on edges and patchwork habitats comprising forests and clear areas.

Distribution: In Sri Lanka, recorded in scrub forests of dry lowlands. Likely to turn up in degraded hill forests.

Voice: Bewildering mix of harsh *kraa kraa* calls, harsh, belling shrieks, metallic shrieks, sweet whistles and piping notes uttered in seemingly random mix at a fast pace.

Status: Uncommon Migrant.

Similar to Black Drongo (see left), with deeply forked tail, but has bright red irides and lacks white rictal spot of Black. Colour more slaty-black in Ashy Drongo versus glossy black in Black.

WHITE-BELLIED DRONGO
Dicrurus caerulescens insularis

Size: 24cm

Habitat: Forest edges, disturbed habitats and village gardens. Visits gardens in Colombo that adjoin open, degraded land or marshland. Generally absent in heavily built-up areas. Feeds on invertebrates taken on the wing or gleaned off plants.

Distribution: Dry and wet zones. The two subspecies between them are found throughout lowlands and mid-hills. Curiously absent from arid south-eastern corner of Sri Lanka, though migrant Ashy Drongos (see p. 270) occupy this area.

Voice: Sometimes pairs utter whinnying call. Wide repertoire of notes mixed with mimicked sequences of Shikra, Jerdon's Leafbird, domestic cats and so on. Notes include harsh notes, whistles and liquid, belling notes.

Status: Common Resident.

Overall black drongo, with dull red eyes and variable white on belly and vent. Dryzone race *insularis* has more white on underparts extending from vent to belly. Wet zone race *leucopygialis* has white restricted to vent, but variations with more white are seen.

GREATER RACKET-TAILED DRONGO
Dicrurus paradiseus ceylonicus

Size: 35cm

Habitat: Riverine forests of dry zone. Dry zone counterpart of Ceylon Crested Drongo.

Distribution: Restricted to dry zone forests.

Voice: Two-noted call where pitch of notes varies. Complex belling call that is repeated.

Status: Uncommon Resident.

'Rackets' on ends of bare outer tail shafts. Naked tail shafts are long in adults. Ceylon Crested Drongos (see p. 272) with bare shafts can be confused with Greater Racket-tailed Drongos, but in the latter lobes at end of tail are turned inwards. In Ceylon Crested they are turned outwards. In the latter, naked shafts are relatively short. Naked shafts in Ceylon Crested caused confusion in the past with Greater Racket-tailed. The identification and consequently distribution of the two species were clarified by local ornithologist Deepal Warakagoda. Crest on forehead prominent in adults, less obvious in immatures.

CEYLON CRESTED DRONGO
Dicrurus lophorhinus ⓔ

Size: 31cm

Habitat: Good-quality forests in wet lowlands. Still present in rainforest pockets such as Bodhinagala.

Distribution: Restricted to lowlands to mid-elevation wet zone forests.

Voice: Lovely repertoire of belling calls. May continue calling at times for a relatively long period. Also a great mimic.

Status: Uncommon Endemic.

Crest on forehead and deeply forked tail with lobes facing outwards. Tail usually without rackets (but see notes under Greater Racket-tailed Drongo, p. 271, for confusion with birds that have bare shafts). Crest, a tuft-like projection on forehead, only obvious at close quarters.

Woodswallows Family *Artamidae*
These long-winged birds are designed for hawking insects in the air. They are thicker set than swallows and martins, and have heavier bills and large, rounded heads. Roosting birds often huddle together.

ASHY WOODSWALLOW (ASHY SWALLOWSHRIKE)
Artamus fuscus

Size: 19cm

Habitat: Mainly in lowlands, hawking for insects over paddy fields and grassland, using vantage points like telegraph wires.

Distribution: Found in lowlands and mid-hills. Becomes scarcer with elevation. Flocks seen in wetlands in suburbs of Colombo, which is surprising as the species is not especially common anywhere.

Voice: Metallic, wheezy, repeated *kreek*. Shorter contact call, *krik*, in flight.

Status: Uncommon Resident.

Bluish-grey head and upperparts. Underparts dirty white. Broad-based, conical bill. Wings extend beyond tail, which is not short but lends the appearance of being short because of the long wings for extended flight. Compact appearance due to stouter build, not slim like the swallows. Bill structure unusual, as it suggests a seed-eating bird like a member of the sparrow family. Hawks in the air and spends more time in the air than bee-eaters, but is less aerial in its habits than swifts or swallows. Woodswallows have clearly found an aerial niche between these two groups.

Crows, jays, magpies and treepies Family *Corvidae*
These are the primates of the bird world; they are highly intelligent, with more brain matter per body mass than other birds. Crows in genus *Corvus* are all black, but relatives such as Ceylon Magpie are very colourful. Lifestyles of the corvids vary – Ceylon Blue Magpies have helpers at the nest, but most crows raise their young as pairs.

CEYLON BLUE MAGPIE
Urocissa ornata

Size: 47cm

Habitat: Restricted to wet zone forests of fairly significant size.

Distribution: Wet zone from lowlands to highlands.

Voice: Wide repertoire of calls, most of which are harsh and grating, some with metallic intonations.

Status: Uncommon Endemic. Vulnerable on IUCN Red List.

Chocolate-brown on head and wings, with red bill, eye-ring and legs set against blue plumage. Tail graduated with white edges and white tip. Unmistakable bird. Birds display cooperative nesting behaviour, with younger birds helping at the nest. In the 1980s this was a difficult bird to see. It is now habituated in Sinharaja and many people enjoy close views of it.

HOUSE CROW
Corvus splendens protegatus

Size: 43cm

Habitat: Has adapted so well to the presence of humans that it is almost entirely absent from habitats that lack a human presence or are human modified.

Distribution: Widespread; in forests replaced by Indian Jungle Crow.

Voice: Harsh *kaaa kaa* that gave rise to the Sinhalese onomatopoeic name '*Kakkaa*'.

Status: Common Resident.

Can be distinguished from bigger Indian Jungle Crow (see p. 274) by grey nape. In fact, one of its older names is Grey-necked Crow. More elegant crow compared with chunky Indian Jungle Crow.

INDIAN JUNGLE CROW
Corvus [macrorhynchos] culminatus

Size: 49cm

Habitat: The name Jungle Crow betrays that this is the crow that can be seen in natural habitats where the presence of humans is not evident. Co-mingles with House Crow in urban areas.

Distribution: Widespread; not as abundant as House Crow.

Voice: Deep-throated, repeated *kraa*. Much deeper than House Crow's.

Status: Resident.

Chunky, uniformly glossy black bird. Large bill only evident in comparison with House Crow (see p. 273), as all crows have heavy bills. This species is often referred to as Large-billed Crow. Lacks grey nape of House Crow.

Organizations

The Sri Lanka Natural History Society (SLNHS). Email: slnhs@lanka.ccom.lk
Founded in 1912, the SLNHS has remained an active, albeit small society with a core membership of enthusiasts and professionals in nature conservation. The SLNHS organizes varied programmes of lectures and presentations for its members. The subject matter of the talks embraces all fields of natural history, including marine life, birds, environmental issues and the recording thereof via photography and other means. It organizes regular field excursions, which include day trips as well as longer excursions with one or more overnight stays.

Field Ornithology Group of Sri Lanka (FOGSL), Department of Zoology, University of Colombo, Colombo 3. Website: www.fogsl.net, Email: fogsl@slt.lk
FOGSL is the Sri Lankan representative of Bird Life International, and is pursuing the goal of becoming a leading local organization for bird study, bird conservation and carrying the conservation message to the public. It has a programme of site visits and lectures throughout the year, and also publishes the *Malkoha* newsletter and other occasional publications. Education is an important activity, and FOGSL uses school visits, exhibitions, workshops and conferences on bird study and conservation to promote its aims.

Wildlife and Nature Protection Society (WNPS), 86 Rajamalwatta Road, Battaramulla. Website: www.wnpssl.org. Email: wnps@sltnet.lk
The WNPS publishes a biannual journal, *Loris* (in English) and *Warana* (in Sinhalese). *Loris* carries a wide variety of articles, ranging from very casual, chatty pieces, to poetry and technical articles. The society also has a reasonably stocked library on ecology and natural history. Various publications, including past copies of *Loris*, are on sale at its offices.

The Young Zoologists' Association of Sri Lanka (YZA), National Zoological Gardens, Dehiwala. Website: www.yzasrilanka.lk. Email: srilankayza@gmail.com
The YZA has nearly 100 school branches and has also set up branch associations. The bulk of its membership consists of schoolchildren and undergraduates, the rest being graduates, professionals and nature lovers from all walks of life.

References

Sri Lankan Wildlife and Related Books

de Silva Wijeyeratne, G. (2015) *A Naturalist's Guide to the Birds of Sri Lanka*. John Beaufoy Publishing: UK.
de Silva Wijeyeratne, G. (2015) *A Naturalist's Guide to the Butterflies & Dragonflies of Sri Lanka*. John Beaufoy Publishing: UK.
de Silva Wijeyeratne, G., Warakagoda, D. & de Zylva, Dr T. S. U. (2000) *A Photographic Guide to the Birds of Sri Lanka*. New Holland: London.
de Silva Wijeyeratne, G. (2007) *A Pictorial Guide and Checklist of the Birds of Sri Lanka*. Jetwing Eco Holidays: Colombo.
de Silva Wijeyeratne, G. (2007) *Sri Lankan Wildlife*. Bradt Travel Guides: UK.
de Silva Wijeyeratne, G. (2013) *Wild Sri Lanka*. John Beaufoy Publishing: UK.
Grimmett, R., Inskipp, C. & Inskipp, T. (2012) *Birds of the Indian Subcontinent*. Helm Field Guides. Christopher Helm: London.
Grimmett, R., Inskipp, C. & Inskipp, T. (1998) *Birds of the Indian Subcontinent*. Christopher Helm: London.
Harrison, J. (1999) *A Field Guide to the Birds of Sri Lanka*. 48 colour plates by Tim Worfolk. Oxford University Press: Oxford.
Henry, G. M. (1998) *A Guide to the Birds of Ceylon*. 3rd revised edn. Oxford University Press: India.
Kotagama, S., & Fernando, P. (1994) *A Field Guide to the Birds of Sri Lanka*. Wildlife Heritage Trust: Colombo.

Warakagoda, D., Inskipp, C., Inskipp, T., & Grimmett, R. (2012) *Birds of Sri Lanka*. Helm Field Guides. Christopher Helm: London.

PAPERS AND OTHER REFERENCES

Cooper, D. & Kay, B. (2009) Field identification of a Pintail Snipe. *BirdingWorld* 22(9), pp. 392–4.

del Hoyo, J., Collar, N. J., Christie, D. A., Elliott, A., Lincoln D. C. & L. D. C. (2014) *Illustrated Checklist of the Birds of theWorld. Volume 1 (Non-passerines)*. HBW and BirdLife International.

De Mel, R. K., Dayanada, S. K., Sumanapala, A., Perera, T., Nanayakkara, R. & Vithanage, M. (2014) A previously undescribed feature in the plumage of juvenile Crested Goshwak *Accipiter tirvirgatues layardi* on Sri Lanka. *BirdingAsia* 21, pp. 100-101.

Dickinson, E. C. (ed.) (2003) *The Howard & Moore Complete Checklist of the Birds of theWorld*. 3rd edn. Christopher Helm: London.

Eck, S. & J. Martens. (2006) Systematic notes on Asian birds. 49. A preliminary review of the Aegithalidae, Remizidae and Paridae. *Zool. Med. Leiden* 80–5(1), 21.xii.

Erritzoe, J., Mann, C. F., Brammer, F. P., & Fuller, R. A. (2012). *Cuckoos of theWorld*. Christopher Helm: London.

Forsman, D. (1999) *Raptors of Europe and the Middle East: A Handbook of Field Identification*. T & D Poyser: London.

Gill, F. & Donsker. D. (eds). (2014) IOC World Bird List (v 4.2). Doi 10.14344/IOC.ML.4.2. www.worldbirdnames.org

IUCN Sri Lanka and the Ministry of Environmental Resources. (2007) *The 2007 Red List of Threatened Fauna and Flora of Sri Lanka*. Colombo, Sri Lanka.

James, D. (2004). Identification of Christmas Island, Great and Lesser Frigatebird. *Birding Asia*. Number I. June 2004. Oriental Bird Club.

Kaluthota, C. (2006) Discovery of a new resident bird species from Sri Lanka. *Siyoth* 1: 45-47-49.

O'Brien, M., Crossely, R., & Karlson, K. (2007). *The Shorebird Guide*. Christopher Helm: London.

Phillips, W. W. A. (1978) *Annotated Checklist of the Birds of Ceylon (Sri Lanka)*. Revised edn. Wildlife and Nature Protection Society of Sri Lanka in Association with the Ceylon Bird Club: Colombo.

Sangster, G., Collinson, J. M., Knox, A. G., Parkin, D. T. & Lars Svensson. Taxonomic recommendations for British birds: Fourth report. *Ibis*. vol. 149, Issue 4, pp. 853–7, October 2007.

Svensson, L., Mullarney, K., Zetterstrom, D. & Grant, P. (2009) *Collins Bird Guide*. 2nd edn. Harper Collins Publishers: London.

Tudge, C. (2008) *The Bird: A Natural History of Who Birds Are, How They Live and Why They Matter*. Crown Publishers: New York.

Vidanapathirana, D. R., Prachnarathna, K. D., Rajeev, M. D. G., & Bandrara, S. (2014) Blue-and-white Flycatcher *Cyanoptila cyanomelana*: first record for Sri Lanka. *Birding Asia* 21 (2014).

Vincombe, K., Harris, A. & Tucker, L. (2014) *The Helm Guide to Bird Identification: An In-depth Look at Confusion Species*. Christopher Helm: London.

Warakagoda, D. (2000) Observations of a Ceylon Crested Drongo with a 'racket tail'. *Ceylon Bird Club Notes*. January: 10–14.

Warakagoda, D. (2001) Colour variation and identification of the male Ceylon Frogmouth (*Batrachostomus moniliger*). *Ceylon Bird Club Notes*. June: 98–104.

Wijesinghe, D. P. (1994) Checklist of the Birds of Sri Lanka. *Ceylon Bird Club Notes Special Publication Series No. 2*, Ceylon Bird Club. Colombo.

Wikramanayake, T. (2015) The first breeding record of Fulvous Whistling-duck *Dendrocygna bicolor* in Sri Lanka. *BirdingAsia* vol. 24. Oriental Bird Club: UK. pp. 90–92.

Zimmerman, D. A., Turner, A. A. & Pearson, D. J. (1999) Reprinted 2005. *Birds of Kenya and Northern Tanzania*. Helm Field Guides. Christopher Helm: London.

FURTHER INFORMATION

There are many detailed articles on birdwatching in Sri Lanka, from overviews, to where to watch seabirds and site-specific accounts. More than 330 articles and a number of publications in pdf format can be copied from a series of Google folders that can be accessed via the author's blog site. See http://wildlifewithgehan.blogspot.co.uk/2016/01/sri-lanka-wildlife-publications.html.

Tour Operators

A strength of Sri Lanka lies in the presence of both general and specialist tour operators that can tailor a birding holiday. A non-exhaustive list of nearly 20 companies that the author is acquainted with is given below.

A Baur & Co. (Travels), www.baurs.com
Adventure Birding, www.adventurebirding.lk
Aitken Spence Travels, www.aitkenspencetravels.com
& Beyond, www.andbeyond.com
Birding Sri Lanka.com www.birdingsrilanka.com
Bird and Wildlife Team, www.birdandwildlifeteam.com
Birdwing Nature Holidays, www.birdwingnature.com
Eco Team (Mahoora Tented Safaris), www.srilankaecotourism.com
Hemtours (Diethelm Travel Sri Lanka), www.hemtours.com
High Elms Travel, www.highelmstravel.com
Jetwing Eco Holidays, www.jetwingeco.com
Lanka Sportreizen, www.lsr-srilanka.com
Little Adventures, www.littleadventuressrilanka.com
Nature Trails, www.naturetrails.lk
Quickshaws Tours, www.quickshaws.com
Red Dot, www.reddottours.com
Sri Lanka in Style, www.srilankainstyle.com
Walkers Tours, www.walkerstours.com
Walk with Jith, www.walkwithjith.com

Acknowledgements

General acknowledgements

Many people have over the years helped me in one way or another to become better acquainted with the natural history of Sri Lanka. My field work has also been supported by several tourism companies as well as state agencies and their staff. To all of them, I am grateful. I must, however, make a special mention of the corporate and field staff of Jetwing Eco Holidays and its sister companies Jetwing Hotels and Jetwing Travels. During my 11 years of residence in Sri Lanka they hugely supported my efforts to draw attention to Sri Lanka as being super-rich in wildlife. Past and present Jetwing Eco Holidays and field staff including Chandrika Maelge, Amila Salgado, Ajanthan Shantiratnam, Paramie Perera, Nadeeshani Attanayake, Ganganath Weerasinghe, Riaz Cader, Ayanthi Samarajewa, Shehani Seneviratne, Aruni Hewage, Divya Martyn, L. S. de S Gunasekera, Chadraguptha Wickremesekera ('Wicky'), Supurna Hettiarachchi ('Hetti'), Chaminda Jayaweera, Sam Caseer, Chandra Jayawardana, Nadeera Weerasinghe, Anoma Alagiyawadu, Hasantha Lokugamage 'Basha', Wijaya Bandara, Suranga Wewegedara, Prashantha Paranagama, Nilantha Kodithuwakku, Dithya Angammana, Asitha Jayaratne, Lal de Silva and various interns have helped me. In the corporate team, Hiran Cooray, Shiromal Cooray, Ruan Samarasinha, Raju Arasaratnam, Sanjiva Gautamadasa and Lalin de Mel and many others have supported my efforts.

My Uncle Dodwell de Silva took me on leopard safaris at the age of three and got me interested in birds. My late Aunt Vijitha de Silva and my sister Manouri got me my first cameras. My late parents Lakshmi and Dalton provided a lot of encouragement; perhaps they saw this as a good way of keeping me out of trouble. My sisters Indira, Manouri, Janani, Rukshan, Dileeni and Yasmin and brother Suraj, also encouraged my pursuit of natural history.

In the UK, my sister Indira and her family always provided a home when I was bridging islands. Dushy and Marnie Ranetunge also helped me greatly on my return to the UK.

My one time neighbour Azly Nazeem, a group of then schoolboys including Jeevan William, Senaka Jayasuriya and Lester Perera and my former scout master Mr. Lokanathan were a key influence in my school days.

My development as a writer is owed to many people. Firstly my mother Lakshmi and more lately various editors who encouraged me to write. A bird book such as this inevitably draws on childhood inspiration from books that were available then. G. M. Henry and W. W. A Phillips, whom I never met, armed me with the literature I needed as a teenage birder. I was also inspired by the work in *Loris* by photographers such as Dr T. S. U. de Zylva and by the programme of events organized by the Wildlife and Nature Protection Society, Field Ornithology Group of Sri Lanka (FOGSL) and the Sri Lanka Natural History Society. The late Thilo Hoffmann of the Ceylon Bird Club and Professor Sarath Kotagama have been inspirations for their tireless efforts with bird conservation. Early mentors in the field included Rex de Silva who took me and a bunch of inexperienced schoolboys under his wing to help with his seabird study. I have also been fortunate to continue having good company in the field in Sri Lanka on my visits after I returned to London. My friends and field companions include Ajith Ratanayaka, Nigel Forbes and Ashan Seneviratne, who have arranged a number of field trips for me.

My wife Nirma and my two daughters Maya and Amali are part of the team. They put up with me not spending the time they deserve with them because I spend my private time working on the 'next book'. Nirma, at times with help from parents Roland and Neela Silva, takes care of many things, allowing me more time to spend on taking natural history to a wider audience.

The list of people and organizations who have helped or influenced me is too long to mention individually, and the people mentioned here are only representative. My apologies to those whom I have not mentioned by name; your support did matter.

SPECIFIC ACKNOWLEDGEMENTS

Tara Wikramanayake helped enormously with much useful feedback, copy-editing early drafts of the text and sense-checking my species accounts. She also compiled the lengths of the birds using as reference the two books on the birds of the Indian Subcontinent by Rasmussen et al and Grimmett et al. Her general all-round support on various tasks also made it easier for me to balance the demands on my time as a newly appointed Chair of the London Bird Club when the demands on my time to complete this book were also peaking. Tara Wikramanayake, Kithsiri Gunawardena and Dr Pathmanath Samaraweera provided useful comments in updating the Checklist of the Birds of Sri Lanka in the *Naturalist's Guide*, which helped with the checklist in this book. Tara kept me informed of new records to ensure the checklist was kept up to date. Any responsibility for errors in ascribing a status remains mine. Many years ago, Avanti Wadugodapitiya helped to 'fit' into a standard structure species text I had written that was used in the *Naturalist's Guide* and which I have continued to use in this book. I have adapted the birding itinerary from a real one used by Jetwing Eco Holidays, developed during my time with Jetwing.

Both John Beaufoy and Rosemary Wilkinson were gently persistent that I should embark on another book with them. My thanks to the photographers who are individually credited, for their images. I once again benefited from Krystyna Mayer, an experienced natural history copy-editor performing the final edit.

PHOTOGRAPHERS

Amy Tsang in Singapore very kindly introduced me to a number of photographers in Singapore to assist me in sourcing images. A large number of photographers have provided me with images and I am grateful to all of them for supporting me. My task in sourcing images was helped greatly by a few photographers having a significant number of the images I needed. These include Tom Tams, Ajith Ratnayaka, Manjula Manthur and Graham Ekins. I am just as grateful to all of the photographers, just over 50 in total, who are listed in the picture credits who made this book possible.

Checklist of the Birds of Sri Lanka

Some species have two or more subspecies (races) recorded in Sri Lanka, in which case the status is given here for each subspecies. For simplicity, in this checklist birds are listed at species level ignoring the trinomials. The presence of more than one status indicates that more than one subspecies or geographical race is present. In the species accounts in the main part of the book, trinomials are provided for many species.

Key to Status

The key used for the islandwide status of the listed species below is as follows. The abundance qualifier on the left combined with the resident/non-resident category indicate the overall status of the bird.

Abundance	Resident/Visitor
C Common	R Resident
U Uncommon	M Migrant
S Scarce	E Endemic
H Highly Scarce	V Vagrant

PODICIPEDIFORMES

Grebes (Podicipedidae)

1	Little Grebe *Tachybaptus ruficollis*	R

PROCELLARIIFORMES

Petrels and shearwaters (Procellariidae)

2	Cape Petrel *Daption capense*	V
3	Barau's Petrel *Pterodroma baraui*	HSM
4	Bulwer's Petrel *Bulweria bulwerii*	V
5	Jouanin's Petrel *Bulweria fallax*	V
6	Streaked Shearwater *Calonectris leucomelas*	V
7	Wedge-tailed Shearwater *Puffinus pacificus*	M
8	Sooty Shearwater *Puffinus griseus*	V
9	Flesh-footed Shearwater *Puffinus carneipes*	M
10	Short-tailed Shearwater *Puffinus tenuirostris*	V
11	Persian Shearwater *Puffinus persicus*	SM

Storm-Petrels (Hydrobatidae)

12	Wilson's Storm-Petrel *Oceanites oceanicus*	HSM, M
13	White-faced Storm-Petrel *Pelagodroma marina*	V
14	Black-bellied Storm-Petrel *Fregetta tropica*	V
15	Swinhoe's Storm-Petrel *Oceanodroma monorhis*	SM

Tropicbirds (Phaethontidae)

16	White-tailed Tropicbird *Phaethon lepturus*	SM
17	Red-billed Tropicbird *Phaethon aethereus*	SM

PELECANIFORMES

Pelicans (Pelecanidae)

18	Spot-billed Pelican *Pelecanus philippensis*	R

Gannets and boobies (Sulidae)

19	Masked Booby Sula dactylatra	SM
20	Brown Booby Sula leucogaster	SM
21	Red-footed Booby Sula sula	V

Cormorants and shags (Phalacrocoracidae)

22	Little Cormorant Phalacrocorax niger	CR
23	Indian Shag Phalacrocorax fuscicollis	CR
24	Great Cormorant Phalacrocorax carbo	SR

Darters (Anhingidae)

25	Oriental Darter Anhinga melanogaster	UR

Frigatebirds (Fregatidae)

26	Lesser Frigatebird Fregata ariel	HSM
27	Great Frigatebird Fregata minor	HSM
28	Christmas Frigatebird Fregata andrewsi	HSM

CICONIIFORMES

Herons and egrets (Ardeidae)

29	Little Egret Egretta garzetta	CR
30	Western Reef-heron Egretta gularis	SM
31	Great Egret Egretta alba	CR
32	Intermediate Egret Egretta intermedia	CR
33	Grey Heron Ardea cinerea	R
34	Goliath Heron Ardea goliath	V
35	Purple Heron Ardea purpurea	R
36	Eastern Cattle Egret Bubulcus coromandus	CR
37	Indian Pond Heron Ardeola grayii	CR
38	Chinese Pond Heron Ardeola bacchus	V
39	Striated Heron Butorides striata	UR
40	Black-crowned Night-heron Nycticorax nycticorax	UR
41	Malayan Night-heron Gorsachius melanolophus	SM
42	Yellow Bittern Ixobrychus sinensis	UR, M
43	Chestnut Bittern Ixobrychus cinnamomeus	UR
44	Black Bittern Dupetor flavicollis	UR, M
45	Eurasian Bittern Botaurus stellaris	V

Storks (Ciconiidae)

46	Painted Stork Mycteria leucocephala	R
47	Asian Openbill Anastomus oscitans	R
48	Black Stork Ciconia nigra	V
49	Woolly-necked Stork Ciconia episcopus	UR
50	White Stork Ciconia ciconia	V
51	Black-necked Stork Ephippiorhynchus asiaticus	HSR
52	Lesser Adjutant Leptoptilos javanicus	SR

Ibises and spoonbills (Threskiornithidae)

53	Glossy Ibis Plegadis falcinellus	SM
54	Black-headed Ibis Threskiornis melanocephalus	R
55	Eurasian Spoonbill Platalea leucorodia	UR

PHOENICOPTERIFORMES

Flamingos (Phoenicopteridae)

56	Greater Flamingo Phoenicopterus roseus	M
57	Lesser Flamingo Phoeniconaias minor	V

ANSERIFORMES

Swans, geese and ducks (Anatidae)

58	Fulvous Whistling-duck Dendrocygna bicolor	V, R
59	Lesser Whistling-duck Dendrocygna javanica	R
60	Greylag Goose Anser anser	V
61	Bar-headed Goose Anser indicus	V
62	Ruddy Shelduck Tadorna ferruginea	V
63	Comb Duck Sarkidiornis melanotos	HSM
64	Cotton Teal Nettapus coromandelianus	UR
65	Gadwall Anas strepera	V
66	Eurasian Wigeon Anas penelope	M
67	Indian Spot-billed Duck Anas poecilorhyncha	HSR, SM
68	Northern Shoveler Anas clypeata	SM
69	Northern Pintail Anas acuta	M
70	Garganey Anas querquedula	M
71	Common Teal Anas crecca	SM
72	Tufted Duck Aythya fuligula	V

FALCONIFORMES

Hawks, kites, eagles and vultures (Accipitridae)

73	Jerdon's Baza Aviceda jerdoni	SR
74	Black Baza Aviceda leuphotes	V
75	Oriental Honey-buzzard Pernis ptilorhyncus	UR, SM
76	Black-winged Kite Elanus caeruleus	UR
77	Black Kite Milvus migrans	SR, M
78	Brahminy Kite Haliastur indus	R
79	White-bellied Sea-eagle Haliaeetus leucogaster	UR
80	Grey-headed Fish-eagle Ichthyophaga ichthyaetus	SR
81	Egyptian Vulture Neophron percnopterus	V
82	Crested Serpent-eagle Spilornis cheela	R
83	Western Marsh Harrier Circus aeruginosus	UM
84	Pallid Harrier Circus macrourus	UM
85	Pied Harrier Circus melanoleucos	V
86	Montagu's Harrier Circus pygargus	UM
87	Crested Goshawk Accipiter trivirgatus	UR
88	Shikra Accipiter badius	R
89	Besra Sparrowhawk Accipiter virgatus	SR
90	Eurasian Sparrowhawk Accipiter nisus	V
91	Himalayan Buzzard Buteo burmanicus	SM
92	Long-legged Buzzard Buteo rufinus	HSM
93	Black Eagle Ictinaetus malayensis	UR
94	Greater Spotted Eagle Clanga clanga	V
95	Tawny Eagle Aquila rapax	V
96	Bonelli's Eagle Hieraaetus fasciatus	V
97	Booted Eagle Hieraaetus pennatus	SM

98	Rufous-bellied Eagle Hieraaetus kienerii	UR
99	Crested Hawk-eagle Spizaetus cirrhatus	R
100	Legge's Hawk-eagle Nisaetus kelaarti	SR

Osprey (Pandionidae)

| 101 | Osprey Pandion haliaetus | SM |

Falcons (Falconidae)

102	Lesser Kestrel Falco naumanni	V
103	Common Kestrel Falco tinnunculus	HSR, UM, HSM
104	Red-headed Falcon Falco chicquera	V
105	Amur Falcon Falco amurensis	HSM
106	Eurasian Hobby Falco subbuteo	V
107	Oriental Hobby Falco severus	V
108	Peregrine Falcon Falco peregrinus	UR,SM

GALLIFORMES

Partridges, quails and pheasants (Phasianidae)

109	Painted Francolin Francolinus pictus	SR
110	Grey Francolin Francolinus pondicerianus	UR
111	Rain Quail Coturnix coromandelica	HSM
112	Blue-breasted Quail Coturnix chinensis	SR
113	Jungle Bush-quail Perdicula asiatica	SR
114	Ceylon Spurfowl Galloperdix bicalcarata	UE
115	Ceylon Junglefowl Gallus lafayetii	CE
116	Indian Peafowl Pavo cristatus	R

GRUIFORMES

Buttonquails (Turnicidae)

| 117 | Small Buttonquail Turnix sylvaticus | V |
| 118 | Barred Buttonquail Turnix suscitator | R |

Rails, crakes, gallinules and coots (Rallidae)

119	Slaty-legged Crake Rallina eurizonoides	UM, HSR
120	Slaty-breasted Rail Rallus striatus	UR, UM
121	Brown-cheeked Water Rail Rallus indicus	V
122	Corn Crake Crex crex	V
123	White-breasted Waterhen Amaurornis phoenicurus	CR
124	'Eastern' Baillon's Crake Porzana pusilla	HSM
125	Ruddy-breasted Crake Porzana fusca	UR
126	Watercock Gallicrex cinerea	UR
127	Purple Swamphen Porphyrio [porphyrio] poliocephalus	R
128	Common Moorhen Gallinula chloropus	R
129	Eurasian Coot Fulica atra	UR,SM

CHARADRIIFORMES

Jacanas (Jacanidae)

| 130 | Pheasant-tailed Jacana Hydrophasianus chirurgus | R |

Painted-snipes (Rostratulidae)

| 131 | Greater Painted-snipe Rostratula benghalensis | UR |

Oystercatchers (Haematopodidae)

132	Eurasian Oystercatcher *Haematopus ostralegus*	SM

Plovers (Charadriidae)

133	Pacific Golden Plover *Pluvialis fulva*	M
134	Grey Plover *Pluvialis squatarola*	UM
135	Common Ringed Plover *Charadrius hiaticula*	SM
136	Little Ringed Plover *Charadrius dubius*	UR, UM
137	Kentish Plover *Charadrius alexandrinus*	UR, UM
138	Lesser Sand Plover *Charadrius mongolus*	M
139	Greater Sand Plover *Charadrius leschenaultii*	SM
140	Caspian Plover *Charadrius asiaticus*	SM
141	Oriental Plover *Charadrius veredus*	V
142	Yellow-wattled Lapwing *Vanellus malabaricus*	UR
143	Grey-headed Lapwing *Vanellus cinereus*	V
144	Red-wattled Lapwing *Vanellus indicus*	CR
145	Sociable Plover *Vanellus gregarius*	V

Sandpipers and allies (Scolopacidae)

146	Eurasian Woodcock *Scolopax rusticola*	HSM
147	Wood Snipe *Gallinago nemoricola*	V
148	Pintail Snipe *Gallinago stenura*	M
149	Swinhoe's Snipe *Gallinago megala*	V
150	Common Snipe *Gallinago gallinago*	UM
151	Great Snipe *Gallinago media*	V
152	Jack Snipe *Lymnocryptes minimus*	HSM
153	Black-tailed Godwit *Limosa limosa*	M, HSM
154	Bar-tailed Godwit *Limosa lapponica*	SM
155	Whimbrel *Numenius phaeopus*	UM
156	Eurasian Curlew *Numenius arquata*	UM
157	Spotted Redshank *Tringa erythropus*	V
158	Common Redshank *Tringa totanus*	CM
159	Common Greenshank *Tringa nebularia*	UM
160	Marsh Sandpiper *Tringa stagnatilis*	CM
161	Green Sandpiper *Tringa ochropus*	UM
162	Wood Sandpiper *Tringa glareola*	M
163	Terek Sandpiper *Xenus cinereus*	UM
164	Common Sandpiper *Actitis hypoleucos*	M
165	Ruddy Turnstone *Arenaria interpres*	M
166	Asian Dowitcher *Limnodromus semipalmatus*	V
167	Great Knot *Calidris tenuirostris*	HSM
168	Red Knot *Calidris canutus*	V
169	Sanderling *Calidris alba*	UM
170	Little Stint *Calidris minuta*	CM
171	Rufous-necked Stint *Calidris ruficollis*	V
172	Temminck's Stint *Calidris temminckii*	SM
173	Long-toed Stint *Calidris subminuta*	SM
174	Sharp-tailed Sandpiper *Calidris acuminata*	V
175	Pectoral Sandpiper *Calidris melanotos*	V
176	Dunlin *Calidris alpina*	V
177	Curlew Sandpiper *Calidris ferruginea*	CM
178	Spoon-billed Sandpiper *Eurynorhynchus pygmeus*	V
179	Buff-breasted Sandpiper *Tryngites subruficollis*	V
180	Broad-billed Sandpiper *Limicola falcinellus*	UM
181	Ruff *Philomachus pugnax*	UM

Ibisbill, stilts and avocets (Recurvirostridae)

182	Black-winged Stilt *Himantopus himantopus*	CR, M
183	Pied Avocet *Recurvirostra avosetta*	SM

Phalaropes (Phalaropodidae)

184	Red-necked Phalarope *Phalaropus lobatus*	SM

Crab-plovers (Dromadidae)

185	Crab-plover *Dromas ardeola*	SR

Stone-curlews and thick-knees (Burhinidae)

186	Indian Stone-curlew *Burhinus indicus*	SR
187	Great Thick-knee *Esacus recurvirostris*	UR

Coursers and pratincoles (Glareolidae)

188	Indian Courser *Cursorius coromandelicus*	HSR
189	Collared Pratincole *Glareola pratincola*	SM
190	Oriental Pratincole *Glareola maldivarum*	UR
191	Small Pratincole *Glareola lactea*	UR

Skuas (Stercorariidae)

192	Brown Skua *Catharacta antarctica*	SM
193	South Polar Skua *Catharacta maccormicki*	V
194	Pomarine Skua *Stercorarius pomarinus*	M
195	Parasitic Skua *Stercorarius parasiticus*	HSM
196	Long-tailed Skua *Stercorarius longicaudus*	V

Gulls (Laridae)

197	Sooty Gull *Larus hemprichii*	V
198	Heuglin's Gull *Larus heuglini*	M
199	'Steppe' Gull *Larus [heuglini] barabensis*	V
200	Great Black-headed Gull *Larus ichthyaetus*	M
201	Brown-headed Gull *Chroicocephalus brunnicephalus*	M
202	Common Black-headed Gull *Chroicocephalus ridibundus*	SM
203	Slender-billed Gull *Larus genei*	V

Terns (Sternidae)

204	Gull-billed Tern *Gelochelidon nilotica*	SR, CM
205	Caspian Tern *Hydroprogne caspia*	SR, M
206	Lesser Crested Tern *Thalasseus bengalensis*	M
207	Great Crested Tern *Thalasseus bergii*	R, V
208	Sandwich Tern *Thalasseus sandvicensis*	HSM
209	Roseate Tern *Sterna dougallii*	UR
210	Black-naped Tern *Sterna sumatrana*	V
211	Common Tern *Sterna hirundo*	SR, SM, M
212	Little Tern *Sterna albifrons*	R
213	Saunders's Tern *Sterna saundersi*	HSR
214	White-cheeked Tern *Sterna repressa*	V
215	Brown-winged (Bridled) Tern *Sterna anaethetus*	M, HSR
216	Sooty Tern *Sterna fuscata*	SM, HSR
217	Whiskered Tern *Chlidonias hybrida*	CM
218	White-winged Tern *Chlidonias leucopterus*	M
219	Brown Noddy *Anous stolidus*	UM, HSR
220	Lesser Noddy *Anous tenuirostris*	HSM

COLUMBIFORMES

Pigeons and doves (Columbidae)

221	Rock Pigeon *Columba livia*	UR
222	Ceylon Woodpigeon *Columba torringtonii*	UE
223	Pale-capped Woodpigeon *Columba punicea*	V
224	Oriental Turtle-dove *Streptopelia orientalis*	HSM
225	Spotted Dove *Streptopelia chinensis*	CR
226	Red Collared-dove *Streptopelia tranquebarica*	V
227	Eurasian Collared-dove *Streptopelia decaocto*	UR
228	Emerald Dove *Chalcophaps indica*	R
229	Orange-breasted Green-pigeon *Treron bicinctus*	R
230	Ceylon Green-pigeon *Treron pompadora*	E
231	Yellow-footed Green-pigeon *Treron phoenicopterus*	HSM, HSR
232	Green Imperial-pigeon *Ducula aenea*	R

PSITTACIFORMES

Parrots (Psittacidae)

233	Ceylon Hanging-parrot *Loriculus beryllinus*	CE
234	Alexandrine Parakeet *Psittacula eupatria*	R
235	Rose-ringed Parakeet *Psittacula krameri*	CR
236	Plum-headed Parakeet *Psittacula cyanocephala*	UR
237	Layard's Parakeet *Psittacula calthropae*	UE

CUCULIFORMES

Cuckoos (Cuculidae)

238	Green-billed Coucal *Centropus chlororhynchos*	USE
239	'Southern' Coucal *Centropus [sinensis] parroti*	CR
240	Sirkeer Malkoha *Taccocua leschenaultii*	UR
241	Red-faced Malkoha *Phaenicophaeus pyrrhocephalus*	SE
242	Blue-faced Malkoha *Phaenicophaeus viridirostris*	UR
243	Chestnut-winged Cuckoo *Clamator coromandus*	UM
244	Jacobin Cuckoo *Clamator jacobinus*	UR
245	Asian Koel *Eudynamys scolopaceus*	CR
246	Asian Emerald Cuckoo *Chrysococcyx maculatus*	V
247	Banded Bay Cuckoo *Cacomantis sonneratii*	UR
248	Grey-bellied Cuckoo *Cacomantis passerinus*	M
249	'Fork-tailed' Drongo-cuckoo *Surniculus [lugubris] dicruroides*	UR
250	Common Hawk-cuckoo *Hierococcyx varius*	UR, HSM
251	Small Cuckoo *Cuculus poliocephalus*	SM
252	Indian Cuckoo *Cuculus micropterus*	UR, M
253	Common Cuckoo *Cuculus canorus*	SM

STRIGIFORMES

Barn-owls and bay owls (Tytonidae)

254	Common Barn-owl *Tyto alba*	SR
255	Ceylon Bay Owl *Phodilus assimilis*	HSR

Owls (Strigidae)

256	Serendib Scops-owl *Otus thilohoffmanni*	SE
257	Oriental Scops-owl *Otus sunia*	SR
258	Indian Scops-owl *Otus bakkamoena*	R
259	Spot-bellied Eagle-owl *Bubo nipalensis*	UR
260	Brown Fish-owl *Ketupa zeylonensis*	R
261	Brown Wood-owl *Strix leptogrammica*	UR
262	Jungle Owlet *Glaucidium radiatum*	R
263	Chestnut-backed Owlet *Glaucidium castanonotum*	E
264	Brown Hawk-owl *Ninox scutulata*	R
265	Short-eared Owl *Asio flammeus*	SM

CAPRIMULGIFORMES

Frogmouths (Podargidae)

266	Ceylon Frogmouth *Batrachostomus moniliger*	R

Nightjars (Caprimulgidae)

267	Great Eared-nightjar *Eurostopodus macrotis*	V
268	Indian Jungle Nightjar *Caprimulgus indicus*	UR
269	Jerdon's Nightjar *Caprimulgus atripennis*	R
270	Indian Little Nightjar *Caprimulgus asiaticus*	R

APODIFORMES

Swifts (Apodidae)

271	Indian Swiftlet *Aerodramus unicolor*	R
272	Brown-throated Needletail *Hirundapus giganteus*	SR
273	Asian Palm-swift *Cypsiurus balasiensis*	R
274	Alpine Swift *Tachymarptis melba*	UR
275	Pacific Swift *Apus pacificus*	V
276	Little Swift *Apus affinis*	R

Treeswifts (Hemiprocnidae)

277	Crested Treeswift *Hemiprocne coronata*	R

TROGONIFORMES

Trogons (Trogonidae)

278	Malabar Trogon *Harpactes fasciatus*	UR

CORACIIFORMES

Kingfishers (Alcedinidae)

279	Common Kingfisher *Alcedo atthis*	R
280	Blue-eared Kingfisher *Alcedo meninting*	HSR
281	Black-backed Dwarf Kingfisher *Ceyx erithaca*	UR
282	Stork-billed Kingfisher *Pelargopsis capensis*	UR
283	White-throated Kingfisher *Halcyon smyrnensis*	CR
284	Black-capped Kingfisher *Halcyon pileata*	SM
285	Lesser Pied Kingfisher *Ceryle rudis*	UR

Bee-eaters (Meropidae)

286	Little Green Bee-eater *Merops orientalis*	R
287	Blue-tailed Bee-eater *Merops philippinus*	CM
288	European Bee-eater *Merops apiaster*	SM
289	Chestnut-headed Bee-eater *Merops leschenaulti*	UR

Rollers (Coraciidae)

290	European Roller *Coracias garrulus*	V
291	Indian Roller *Coracias benghalensis*	R
292	Dollarbird *Eurystomus orientalis*	HSR

Hoopoes (Upupidae)

293	Common Hoopoe *Upupa epops*	UR

Hornbills (Bucerotidae)

294	Ceylon Grey Hornbill *Ocyceros gingalensis*	E
295	Malabar Pied Hornbill *Anthracoceros coronatus*	R

PICIFORMES

Barbets (Capitonidae)

296	Brown-headed Barbet *Megalaima zeylanica*	CR
297	Yellow-fronted Barbet *Megalaima flavifrons*	E
298	Ceylon Small Barbet *Megalaima rubricapillus*	CE
299	Coppersmith Barbet *Megalaima haemacephala*	R

Woodpeckers (Picidae)

300	Eurasian Wryneck *Jynx torquilla*	HSM
301	Brown-capped Pygmy Woodpecker *Dendrocopos nanus*	UR
302	Yellow-fronted Pied Woodpecker *Dendrocopos mahrattensis*	UR
303	Rufous Woodpecker *Micropternus brachyurus*	UR
304	Lesser Yellownape *Picus chlorolophus*	R
305	Streak-throated Woodpecker *Picus xanthopygaeus*	SR
306	Black-rumped Flameback *Dinopium benghalense*	UR
307	Ceylon Red-backed Woodpecker *Dinopium psarodes*	CE
308	Crimson-backed Flameback *Chrysocolaptes stricklandi*	E
309	White-naped Flameback *Chrysocolaptes festivus*	SR

PASSERIFORMES

Pittas (Pittidae)

310	Indian Pitta *Pitta brachyura*	M

Larks (Alaudidae)

311	Jerdon's Bushlark *Mirafra affinis*	R
312	Ashy-crowned Finch-lark *Eremopterix griseus*	UR
313	Greater Short-toed Lark *Calandrella brachydactyla*	V
314	Oriental Skylark *Alauda gulgula*	UR, SR

Swallows and martins (Hirundinidae)

315	Common Sand-martin *Riparia riparia*	SM
316	Dusky Crag-martin *Ptyonoprogne concolor*	V
317	Barn Swallow *Hirundo rustica*	CM, UM, SM
318	Hill Swallow *Hirundo domicola*	R

319	Wire-tailed Swallow *Hirundo smithii*	V
320	Red-rumped Swallow *Hirundo daurica*	SM, HSM
321	Ceylon Swallow *Hirundo hyperythra*	E
322	Streak-throated Swallow *Hirundo fluvicola*	V

Pipits and wagtails (Motacillidae)

323	Forest Wagtail *Dendronanthus indicus*	M
324	White Wagtail *Motacilla alba*	SM, HSM
325	White-browed Wagtail *Motacilla maderaspatensis*	V
326	Citrine Wagtail *Motacilla citreola*	HSM
327	Western Yellow Wagtail *Motacilla flava*	M, SM, HSM
328	Grey Wagtail *Motacilla cinerea*	M
329	Richard's Pipit *Anthus richardi*	SM
330	Paddyfield Pipit *Anthus rufulus*	R
331	Tawny Pipit *Anthus campestris*	V
332	Blyth's Pipit *Anthus godlewskii*	SM
333	Olive-backed Pipit *Anthus hodgsoni*	V
334	Red-throated Pipit *Anthus cervinus*	V

Cuckooshrikes (Campephagidae)

335	Large Cuckooshrike *Coracina macei*	UR
336	Black-headed Cuckooshrike *Coracina melanoptera*	R
337	Small Minivet *Pericrocotus cinnamomeus*	R
338	Orange Minivet *Pericrocotus flammeus*	R
339	Pied Flycatcher-shrike *Hemipus picatus*	UR
340	Ceylon Woodshrike *Tephrodornis affinis*	E

Monarchs (Monarchidae)

| 341 | Asian Paradise Flycatcher *Terpsiphone paradisi* | R, M |
| 342 | Black-naped Blue Monarch *Hypothymis azurea* | UR |

Fantails (Rhipiduridae)

| 343 | White-browed Fantail *Rhipidura aureola* | R |

Bulbuls (Pycnonotidae)

344	Black-capped Bulbul *Pycnonotus melanicterus*	UE
345	Red-vented Bulbul *Pycnonotus cafer*	CR
346	Yellow-eared Bulbul *Pycnonotus penicillatus*	E
347	White-browed Bulbul *Pycnonotus luteolus*	R
348	Yellow-browed Bulbul *Iole indica*	UR
349	Square-tailed Black Bulbul *Hypsipetes ganeesa*	UR

Ioras (Aegithinidae)

| 350 | Common Iora *Aegithina tiphia* | R |
| 351 | Marshall's Iora *Aegithina nigrolutea* | SR |

Leafbirds (Chloropseidae)

| 352 | Gold-fronted Leafbird *Chloropsis aurifrons* | UR |
| 353 | Jerdon's Leafbird *Chloropsis jerdoni* | R |

Shrikes (Laniidae)

354	Brown Shrike *Lanius cristatus*	CM, SM
355	Bay-backed Shrike *Lanius vittatus*	V
356	Long-tailed Shrike *Lanius schach*	SR
357	Southern Grey Shrike *Lanius meridionalis*	V

Thrushes (Turdidae)

358	Pied Ground-thrush *Zoothera wardii*	UM
359	Orange-headed Thrush *Zoothera citrina*	SM
360	Spot-winged Ground-thrush *Zoothera spiloptera*	E
361	Ceylon Scaly Thrush *Zoothera imbricata*	SE
362	Indian Blackbird *Turdus simillimus*	R
363	Eyebrowed Thrush *Turdus obscurus*	V
364	Ceylon Whistling-thrush *Myophonus blighi*	SE

Old World flycatchers and chats (Muscicapidae)

365	Asian Brown Flycatcher *Muscicapa dauurica*	M
366	Brown-breasted Flycatcher *Muscicapa muttui*	M
367	Yellow-rumped Flycatcher *Ficedula zanthopygia*	V
368	Kashmir Flycatcher *Ficedula subrubra*	SM
369	Dusky Blue Flycatcher *Eumyias sordidus*	UE
370	Blue-and-White Flycatcher *Cyanoptila cyanomelana*	V
371	Blue-throated Flycatcher *Cyornis rubeculoides*	V
372	Tickell's Blue Flycatcher *Cyornis tickelliae*	R
373	Grey-headed Canary-flycatcher *Culicicapa ceylonensis*	UR
374	Blue Rock-thrush *Monticola solitarius*	SM
375	Bluethroat *Luscinia svecica*	V
376	Indian Blue Robin *Luscinia brunnea*	M
377	Rufous-tailed Scrub-robin *Cercotrichas galactotes*	V
378	Oriental Magpie-robin *Copsychus saularis*	CR
379	White-rumped Shama *Copsychus malabaricus*	R
380	Indian Black Robin *Saxicoloides fulicatus*	R
381	Whinchat *Saxicola rubetra*	V
382	Siberian Stonechat *Saxicola maurus*	V
383	Pied Bushchat *Saxicola caprata*	UR
384	Pied Wheatear *Oenanthe pleschanka*	V
385	Desert Wheatear *Oenanthe deserti*	V
386	Isabelline Wheatear *Oenanthe isabellina*	V

Babblers (Timaliidae)

387	Ashy-headed Laughingthrush *Garrulax cinereifrons*	SE
388	Brown-capped Babbler *Pellorneum fuscocapillus*	UE
389	Ceylon Scimitar-babbler *Pomatorhinus [schisticeps] melanurus*	UE
390	Tawny-bellied Babbler *Dumetia hyperythra*	UR
391	Dark-fronted Babbler *Rhopocichla atriceps*	R
392	Yellow-eyed Babbler *Chrysomma sinense*	R
393	Ceylon Rufous Babbler *Turdoides rufescens*	UE
394	Yellow-billed Babbler *Turdoides affinis*	CR

Cisticolas, prinias and tailorbirds (Cisticolidae)

395	Zitting Cisticola *Cisticola juncidis*	R
396	Grey-breasted Prinia *Prinia hodgsonii*	UR
397	Ashy Prinia *Prinia socialis*	R
398	Jungle Prinia *Prinia sylvatica*	R
399	Plain Prinia *Prinia inornata*	R
400	Common Tailorbird *Orthotomus sutorius*	CR

Old World warblers (Sylviidae)

401	Sri Lanka Bush-warbler *Elaphrornis palliseri*	UE
402	Lanceolated Warbler *Locustella lanceolata*	V
403	Grasshopper Warbler *Locustella naevia*	V
404	Rusty-rumped Warbler *Locustella certhiola*	SM

405	Blyth's Reed-warbler *Acrocephalus dumetorum*	CM
406	Indian Reed-warbler *Acrocephalus [stentoreus] brunnescens*	UR
407	Sykes's Warbler *Hippolais rama*	SM
408	Booted Warbler *Hippolais caligata*	V
409	Dusky Warbler *Phylloscopus fuscatus*	V
410	Greenish Warbler *Phylloscopus trochiloides*	SM
411	Bright-green Warbler *Phylloscopus nitidus*	CM
412	Large-billed Leaf-warbler *Phylloscopus magnirostris*	CM
413	Western Crowned Warbler *Phylloscopus occipitalis*	V
414	Indian Broad-tailed Grass-warbler *Schoenicola platyurus*	V
415	Lesser Whitethroat *Sylvia curruca*	V
416	'Desert' Whitethroat *Sylvia [curruca] minula*	V
417	Hume's Whitethroat *Sylvia althaea*	SM

Titmice (Paridae)

418	Cinereous Tit *Parus cinereous*	R

Nuthatches (Sittidae)

419	Velvet-fronted Nuthatch *Sitta frontalis*	R

Flowerpeckers (Dicaeidae)

420	Thick-billed Flowerpecker *Dicaeum agile*	UR
421	Legge's Flowerpecker *Dicaeum vincens*	UE
422	Pale-billed Flowerpecker *Dicaeum erythrorhynchos*	CR

Sunbirds (Nectariniidae)

423	Purple-rumped Sunbird *Leptocoma zeylonica*	CR
424	Purple Sunbird *Cinnyris asiaticus*	R
425	Loten's Sunbird *Cinnyris lotenius*	R

White-eyes (Zosteropidae)

426	Ceylon White-eye *Zosterops ceylonensis*	E
427	Oriental White-eye *Zosterops palpebrosus*	R

Buntings (Emberizidae)

428	Grey-necked Bunting *Emberiza buchanani*	V
429	Black-headed Bunting *Emberiza melanocephala*	V
430	Red-headed Bunting *Emberiza bruniceps*	V

Finches (Fringillidae)

431	Common Rosefinch *Carpodacus erythrinus*	V

Waxbills (Estrildidae)

432	Indian Silverbill *Euodice malabarica*	UR
433	White-rumped Munia *Lonchura striata*	R
434	Black-throated Munia *Lonchura kelaarti*	UR
435	Scaly-breasted Munia *Lonchura punctulata*	R
436	Tricoloured Munia *Lonchura malacca*	R

Old World sparrows (Passeridae)

437	House Sparrow *Passer domesticus*	R
438	Yellow-throated Sparrow *Petronia xanthocollis*	V

Weavers (Ploceidae)

439	Streaked Weaver *Ploceus manyar*	UR
440	Baya Weaver *Ploceus philippinus*	R

Starlings and mynas (Sturnidae)

441	White-faced Starling *Sturnornis albofrontatus*	HSE
442	Grey-headed Starling *Sturnia malabarica*	V
443	Daurian Starling *Sturnia sturnina*	V
444	Brahminy Starling *Temenuchus pagodarum*	M
445	Rosy Starling *Sturnus roseus*	M
446	Common Myna *Acridotheres tristis*	CR
447	Ceylon Hill-myna *Gracula ptilogenys*	UE
448	Lesser Hill-myna *Gracula indica*	R

Orioles (Oriolidae)

449	European Golden Oriole *Oriolus oriolus*	SM
450	Indian Golden Oriole *Oriolus kundoo*	SM
451	Black-naped Oriole *Oriolus chinensis*	V
452	Black-hooded Oriole *Oriolus xanthornus*	R

Fairy-bluebirds (Irenidae)

453	Asian Fairy-bluebird *Irena puella*	V

Drongos (Dicruridae)

454	Black Drongo *Dicrurus macrocercus*	UR
455	Ashy Drongo *Dicrurus leucophaeus*	UM
456	White-bellied Drongo *Dicrurus caerulescens*	CR
457	Greater Racket-tailed Drongo *Dicrurus paradiseus*	UR
458	Ceylon Crested Drongo *Dicrurus lophorinus*	UE

Woodswallows (Artamidae)

459	Ashy Woodswallow *Artamus fuscus*	UR

Crows, jays, magpies and treepies (Corvidae)

460	Ceylon Blue Magpie *Urocissa ornata*	UE
461	House Crow *Corvus splendens*	CR
462	Indian Jungle Crow *Corvus* [*macrorhynchos*] *culminatus*	R

Index